CRITICAL ENVIRONMENTS: NATURE, SCIENCE, AND POLITICS

*Edited by Julie Guthman and Rebecca Lave*

The Critical Environments series publishes books that explore the political forms of life and the ecologies that emerge from histories of capitalism, militarism, racism, colonialism, and more.

*The Pathogens of Finance*

# The Pathogens of Finance

HOW CAPITALISM BREEDS VECTOR-
BORNE DISEASE

Brent Z. Kaup and
Kelly F. Austin

UNIVERSITY OF CALIFORNIA PRESS

University of California Press
Oakland, California

© 2025 by Kelly Noble and Brent Z. Kaup

Cataloging-in-Publication Data is on file at the Library of Congress.

ISBN 978-0-520-41249-1 (cloth)
ISBN 978-0-520-41250-7 (pbk)
ISBN 978-0-520-41251-4 (ebook)


GPSR Authorized Representative: Easy Access System Europe,
Mustamäe tee 50, 10621 Tallinn, Estonia, gpsr.requests@easproject.com

# Contents

# Acknowledgments

This book is both about and a product of infectious diseases. While the two of us had sat in the same rooms at the same conferences prior to the outbreak of COVID-19 in 2019, we never actually spoke face-to-face in the same physical space prior to completing this book. The collaborative conceptualization and writing of the book occurred completely in the virtual world. In addition, our initial presentation of the ideas underpinning this book occurred at the 2022 annual conference of the Association of American Geographers, which was forced to go virtual due to a resurgent wave of COVID-19.

As academics with children, we wrote this book with what precious little time we could carve out of our pandemic-afflicted schedules. From spring 2020 through summer 2021, Brent helped oversee his kids' education and an afternoon pod when the local schools went virtual. During this time, he frequently contemplated how many mosquitoes were breeding in his downspouts while watching hour-long kindergarten show-and-tells over Zoom. Lacking childcare and seeking to enhance his then–second graders' science education, he also brought his masked kid into the field to collect ticks with the students in his college class on tick-borne disease. Kelly took turns with her husband, trading off working and taking care of their kids, who were both under two years old when the pandemic first hit. She spent many hours hiding out in her partially finished attic, trying to focus on writing while the sounds of toddlers playing echoed nearby. Despite being extra careful and persistent mask wearers in the early years of the pandemic, both Brent and Kelly caught COVID during the writing of this book. Brent and his family experienced two COVID Christmases in a row. Kelly and her family caught COVID late in the pandemic and were repeatedly plagued by other illnesses that the children of quarantine seem to more easily catch and spread. While all of us recovered from the physical manifestations of COVID-19, the pandemic definitely delayed this book's creation. In addition, as scholars who study infectious diseases, we know firsthand that the guilt of catching one always weighs on the human conscience.

Like most books and academic undertakings, we owe too many people thanks. Brent first started his research on ticks and tick-borne disease after Matthias Leu and Oliver Kerscher convinced him to help form the Socionatural Tick-Borne Disease Research Group (STBDRG). Through this interdisciplinary collaboration, Brent learned more about tick ecology and tick-borne disease genetics than he ever imagined he would need or care to know. He also still has ticks in vials of alcohol stored in his freezer that he collected off his clothes for the project. Too many students to name here worked with Brent as part of the group, all of whom deserve thanks. But of particular note, Matthew Abel and Amanda Sikirica helped push Brent to think more about the impact of and responses to the tick-borne disease. Likely the STBDRG's best tick collector, Julia Moore also helped Brent more fully appreciate how walking through the woods to collect ticks could be conceptualized as a stress relief activity. Mark Cooper deserves a note of thanks as well. Working on the project as the William & Mary Environmental Science and Policy Program Mellon Postdoctoral Fellow, Mark (perhaps now begrudgingly) did fieldwork for the STBDRG during one of the hottest Virginia summers on record. He and his partner, Jane (who sometimes helped with such activities), had the pleasure of stepping in a freshly hatched batch of tick larva on one outing. I am sure both never want to think about ticks again. Mark also served as an excellent sounding board of sorts for some of the early theoretical ideas undergirding this book. Brent further owes thanks to the numerous scientists, policy makers, and activists who helped him to better understand both the ecological conditions in which emerging infectious diseases are more likely to occur and the local responses to them.

Kelly became inspired to study coffee after spending a few summers in Bududa and staying with the Zaale family. None of her research would have been possible without their help. In particular, Kelly owes a debt of gratitude to Dezz Natala Zaale. Dezz was instrumental in translating dozens of interviews, helping to find coffee farmers and traders to talk to, and always being willing to lend a hand (sometimes literally when helping Kelly scale steep slopes to hike to coffee farms). Dezz and his family, including his parents, David and Elizabeth Zaale, spent countless hours talking with Kelly about coffee, Uganda's history, and their life experiences, all of which greatly influenced and informed the research. Doctor James Wafula also deserves much thanks. He spent hundreds of hours mentoring Kelly in clinical settings as they worked together to provide medical care to malaria patients. James taught Kelly all about malaria, including how to identify the parasite in human blood using microscopy. Kelly also received help from numerous other partners and community members in Uganda. Over the years, the openness and warmth of the Bududa community has allowed her to share many beautiful memories with the people she has interviewed and worked with. Kelly also wants to acknowledge the work done by her students. Zach Sokol and Ginger Berndt helped to transcribe many of the interviews used in this book. In addition, countless other students who have taken part in Lehigh University's Uganda program over the years have supported Kelly's research in direct and indirect ways.

Both of us are appreciative of everyone involved in the University of California Press's Critical Environments series. Samer Alatout kindly read an initial draft of our book proposal, provided us with useful suggestions, and introduced us to the series editors. Julie Guthman's enthusiasm for the book and excellent suggestions in the initial round of reviews both kept us motivated and helped us to put together a better manuscript. Julie truly understood our project from the beginning, and her comments and ideas helped us want to make the book more accessible to a broader audience. Naja Pulliam Collins's editorial guidance and advice has been incredibly useful. Like Julie, Naja seemed to get what we were trying to do from day one. In addition, her suggestions along the way helped us to stream-line and clarify several parts of the book. We also thank Sylvie Bower for her help and guidance in getting all the appropriate documents together in the right formats and at the right times and Steven Baker for careful and close copyediting.

Mark Noble and Amy Quark deserve thanks for reading and providing feedback on various drafts of this manuscript. Their ideas and close reads helped us clarify several parts of the book. Our departmental colleagues and administrators also deserve thanks. As our careers have progressed, we have come to realize that working in collegial departments is more of an exception than the norm. We are thus grateful for the kindness and support we have received from our colleagues over the years. Finally, we owe thanks to our partners and parents for providing support and aiding with childcare during this book's creation. Given the already mentioned trials and tribulations of writing a book in part during a pandemic, finishing would likely have taken twice as long without such help.

Portions of Brent's research were funded by the William & Mary Commonwealth Center for Energy and Environment, a William & Mary Faculty Research Grant, and the William & Mary W. Taylor Reveley III Interdisciplinary Faculty Fellowship. Portions of Kelly's research were funded by Lehigh University, under its Faculty Research Grants awarded by the Office of the Vice Provost for Research. Chapters 2, 3, and 5 of this book are derived in part from articles published in *Environmental Sociology*, *Antipode*, and *Environment and Planning E*.[1]

The cover art for this book was derived from Louis M. Glackens' *To swat 'em is a waste of time*, originally appearing in *Puck* in 1912.

# Abbreviations

| | |
|---|---|
| ACT | artemisinin combination therapy |
| AHCPR | Agency for Health Care Policy and Research |
| Anvisa | Agência Nacional de Vigilância Sanitária (National Health Regulatory Agency) |
| BCU | Bugisu Cooperative Union |
| BNH | Banco Nacional da Habitação (National Housing Bank) |
| CDA | Colonial Development Act |
| CDC | Centers for Disease Control and Prevention |
| CIA | Central Intelligence Agency |
| COVID-19 | Coronavirus Disease 2019 |
| CTNBio | Comissão Técnica Nacional de Biossegurança (National Technical Commission on Biosafety) |
| DDT | dichlorodiphenyltrichloroethane |
| EM | erythema migrans |
| ERISA | Employee Retirement Income Security Act |
| FHA | Federal Housing Administration |

| | |
|---|---|
| FHFA | Federal Housing Finance Agency |
| Funasa | Fundação Nacional de Saúde (National Health Foundation) |
| GMO | genetically modified organism |
| H1N1 | swine influenza |
| HMO | health maintenance organization |
| HOLC | Home Owners Loan Corporation |
| ICA | International Coffee Agreement |
| IDSA | Infectious Disease Society of America |
| IgG | Immunoglobulin G |
| IgM | Immunoglobulin M |
| IMF | International Monetary Fund |
| INCRA | Instituto Nacional de Colonização e Reforma Agrária (National Institute for Colonization and Agrarian Reform |
| JBUSEDC | Joint Brazil United States Economic Development Commission |
| LMO | living modified organism |
| MCO | managed care organization |
| PSRO | Professional Standards Review Organization |
| S&L | savings and loan association |
| SAP | structural adjustment program |
| SEC | Securities and Exchange Commission |
| SIT | sterile insect technique |
| TRIP | trade-related intellectual property right |
| UCDA | Ugandan Coffee Development Authority |
| UN | United Nations |
| WHO | World Health Organization |

# Introduction

In January 2021, Eric O'Keefe published an article in *The Land Report* about a 14,500-acre plot of land named 100 Circle Farms.[1] The land was located in the Columbia River Basin in Benton County, Washington. While the soils in the river basin are some of the most fertile in the United States, the climate in Eastern Washington is extremely arid. However, given that the Columbia River is fed by an abundance of glacial melt, the land surrounding it is easily irrigated. The crops at 100 Circle Farms were watered with central pivot irrigation, forming near-perfect circles on the ground. According to McDonald's, which purchases the potatoes grown on the land to make them into French fries, the circles can be seen from outer space.[2]

While O'Keefe recognized the natural value of the land, it was the recent sale of 100 Circle Farms that drew his interest. In 2018, the land sold for US$1.7 million, nearly US$12,000 an acre.[3] The local newspaper, the *Tri-City Herald*, reported that the land had been bought by a limited liability company associated with Angelina Agriculture, based in Monterrey, Louisiana. However, O'Keefe had never heard of Angelina Agriculture, and the company was not associated with any major landowners in the United States. After some sleuthing, O'Keefe found that

Angelina Agriculture was the name of a farm in Louisiana whose title had been held by a larger real estate investment trust (REIT). In 2017, the REIT was purchased by Cascade Investment LLC, an American holding company that manages the assets of the cofounder and former CEO of Microsoft, Bill Gates. In the sale of 100 Circle, Angelina Agriculture thus served as a front company. According to O'Keefe, Cascade Investment's purchase of 100 Circle Farms helped make Gates the largest owner of farmland in the United States.[4]

Land is one of many financial assets that Gates holds. After cofounding Microsoft in 1975, he held 45% of the shares in the company for close to two decades. In 1995, he began to work with Cascade Investment to diversify his personal investment portfolio away from Microsoft and create an endowment to fund the Bill & Melinda Gates Foundation.[5] For most of the past decade, close to 50% of Cascade Investment's portfolio has been invested in shares of Berkshire Hathaway, a multinational conglomerate holding company. In addition, Cascade Investment holds sizable shares in CN (which owns Canadian National Railway), WM (Waste Management), John Deere, and numerous other companies.[6] At the end of 2023, the Bill & Melinda Gates Foundation Trust, which holds the foundation's endowment, controlled over US$75 billion in assets.[7]

Over the past twenty-five years, the Gates Foundation has used the earnings from its investments to provide financial support to a wide array of health initiatives. Its mission is "to create a world where every person has the chance to live a healthy and productive life."[8] The Gates Foundation started funding health-related research in its early years of existence but deepened its commitment in 2003 when it provided over US$450 million for its Grand Challenges in Global Health.[9] In 2005, the foundation substantially increased its efforts to combat mosquito-borne disease by committing US$258 million to fight malaria.[10] Two years later, Gates publicly declared his foundation would help lead the world in eradicating malaria. Since then, the Gates Foundation has provided hundreds of millions of dollars toward research and targeted health interventions intended to stop the spread of mosquito- and other insect-borne diseases.[11]

Gates and his foundation's actions are, on the surface, highly laudable, but the means through which he believes everyone can have "the chance to live a healthy and productive life" are embedded in a distinct worldview.

Gates's vision depends on the profits and prophets of Wall Street. As for many of the richest people in the world, much of Gates's wealth and the funding for his philanthropic activities rest in finance. His wealth did not merely originate from the creation and sale of a software program, but from the perceived value and sale of Microsoft stock. He has continued to accrue billions of dollars in wealth for himself and the Gates Foundation through the purchase and sale of an array of financial assets. Furthermore, Gates's vision rests on using market mechanisms to correct market failures. While some of the innovations funded by the Gates Foundation have been used and distributed at little to no cost to recipients, Gates has adamantly supported the patenting and sale of many innovations, including some developed with the foundation's help that are intended to improve health outcomes.[12]

The Bill & Melinda Gates Foundation is thus a creature of Wall Street, dependent on the creation of shareholder value without serious regard to how its investments may help create environments more ideal for infectious disease outbreaks, thus undermining the foundation's own attempts to resolve them. Agricultural land, such as that of 100 Circle Farms, has been made into a profitable financial asset only through the massive transformation of the local environment. While no known infectious diseases have emerged from this landscape, on which potatoes for McDonald's fries are grown, such ecological transformations are known to alter local ecologies in ways that enhance the presence of certain insects and the diseases they carry. The Gates Foundation's attempts to alleviate and eradicate such diseases by using the profits from its financial assets to support the development of commercially available pharmaceutical cures and biotech innovations thus do little to address the ecological roots of these problems.

That said, Gates is not alone in perpetuating the fallacy that finance will fund and fix the problems surrounding new and old infectious diseases. A similar belief underpins the actions of like-minded billionaires and their philanthropic organizations, major international lending and development agencies and their government backers, and even the everyday investor seeking to put their money in so-called sustainable forms of growth through the purchase of a share in an environmental, social, and governance (ESG) mutual fund. The myth that finance can save the world has been constructed and ingrained in nearly every part of our lives for decades.

## THE SCOURGE OF INFECTIOUS DISEASE

The scourge of infectious disease has reemerged.[13] The outbreak of the coronavirus (COVID-19) made this fact readily apparent to nearly everyone across the globe. As the virus rapidly spread in 2020, images of sick and dying people filled social media feeds and appeared on the front pages of mainstream newspapers. Within weeks, hospitals and morgues were packed with human bodies that could not be treated or even laid to rest in any semblance of a timely fashion. From 2020 to 2023, nearly 7 million people were reported to have died from the virus, with some estimates placing the death toll closer to 18 million. The highest rates of mortality from COVID-19 were reported in countries from the global south. However, the United States led the per country death toll by sheer numbers, topping 1 million deaths from COVID-19 by early 2023.[14]

While infectious diseases have long plagued human societies, advancements in the early and mid-twentieth century decreased infectious disease incidence and mortality rates. In urban areas, particularly in the global north, improvements to water, sewage, and sanitation infrastructure resulted in noticeable decreases in many infectious diseases.[15] Access to cleaner water decreased rates of bacterial illnesses such as cholera and E. coli. In addition, efforts to control the amount of standing water in heavily populated areas helped decrease mosquito populations and, in turn, outbreaks of yellow fever and malaria. Medical discoveries further decreased the outbreaks and impact of infectious disease. A yellow fever vaccine lowered rates of infection from the virus worldwide, and the discovery of penicillin and other antibiotics provided a cure for bacterial diseases such as tuberculosis and cholera.[16] The development of modern synthetic insecticides such as dichlorodiphenyltrichloroethane (DDT) and their widespread use also resulted in a reduction in vector-borne diseases such as malaria, dengue, and yellow fever.[17] With such infrastructural and technological advancements abating many infectious diseases, the widespread belief emerged that economic development and scientific advancements would eliminate the threat of infectious disease throughout the world.[18] However, during the last decades of the twentieth century, the seeming success of these purported solutions began to falter.

COVID-19 is just one of many infectious diseases that have emerged or spread to plague society over the past forty years. Among the deadliest, the human immunodeficiency virus (HIV) has taken the lives of over 40 million people since it first appeared in the 1980s.[19] Other, newly emerged infectious diseases, such as severe acute respiratory syndrome (SARS), the swine flu (H1N1), and Ebola virus have resulted in fewer deaths but have still taken the lives of tens of thousands of people. And, while less fatal, diseases such as Lyme have disrupted the lives of hundreds of thousands of people.[20] Older infectious diseases have also reemerged and spread to new locations. Infection and mortality rates from both cholera and tuberculosis have seen steady increases over the past twenty years.[21] The Zika virus, historically isolated to West Africa, spread to the Americas in 2015 and was determined to be the cause of microcephaly and other neurological impairments in newborns.[22] Yellow fever outbreaks continue to impact equatorial regions despite having been eradicated in much of the world.[23] And, after decreasing at the beginning of the twenty-first century, malaria rates and deaths are once again rising, annually affecting nearly 250 million people and taking more than 500,000 lives.[24]

Natural scientists often attribute the contemporary increases in infectious disease outbreaks to the alteration of the environments in which we live. An ever-rising number of studies demonstrate that human-induced climate change has both increased the frequency and intensity of infectious disease outbreaks and expanded the geographic ranges in which an array of infectious diseases can be found.[25] In 2022, a group of scientists found that 58% of known human pathogenic diseases can be aggravated by climate change.[26] As average temperatures rise and moisture levels change, the distribution and abundance of disease-carrying animals and insects also expand. Warmer temperatures allow such creatures to more easily migrate and live farther away from equatorial regions. In addition, surges in rainfall can overwhelm existing sewage and sanitation systems, increasing the presence of standing water where disease carrying insects can in turn lay their eggs.

Natural scientists have also linked anthropogenic landscape change to the proliferation of infectious diseases.[27] In rural areas, turning forests into agricultural fields is known to create more ideal opportunities for infectious disease spillovers. On a global level, the total area of land covered by forests has decreased by 81.7 million hectares since 1960.[28] While forest cover has increased in some parts of the world, deforestation in equatorial regions where diseases such as yellow fever, dengue, and malaria are more prevalent has outpaced such efforts. As tropical forests are converted into row crops, livestock pastures, and palm oil plantations, the people in such places and the domesticated animals they tend are more likely to come into contact with wildlife and insects that can carry an array of infectious diseases. At the same time, the destruction of forest ecosystems dilutes biodiversity and in turn potentially shrinks populations of the predators and competitors of pathogen-carrying wildlife and insects.[29] The changing landscapes in urban areas have also created spaces more conducive to infectious disease outbreaks.[30] Investments in water, sewage, and sanitation infrastructure have often failed to keep up with population growth, particularly in more impoverished parts of the world. In addition, much of the world's existing water, sewage, and sanitation infrastructure has aged past its effective use date and is slowly degrading.[31] A lack of access to clean water and absence of adequate sewage and sanitation systems have been shown to increase the risk of both bacterial illnesses such as E. coli and cholera and vector-borne diseases such as yellow fever and dengue.

As changes to our planetary and local ecologies have increased the likelihood of infectious disease outbreaks, the twentieth-century interventions used to treat infectious diseases and reduce the presence of the creatures that carry them have become less effective. Natural scientists and public health scholars have raised the alarm about antibiotic, antimicrobial, and insecticide resistance. Bacteria, viruses, and protozoa that cause infectious diseases that could once be treated with medicines have evolved.[32] So too have the insects carrying infectious diseases that could once be killed by insecticides.[33] Perhaps most notably, the malaria parasite has developed resistance to all known treatments. While antimalarial medicines still largely work, the resistant strains of the parasite will likely continue to strengthen and spread, making the treatment of malaria difficult if not impossible.[34] At the same time, the mosquitoes carrying malaria have developed resistance to conventional insecticides, including pyrethroids, the dominant class of chemicals used on insecticide-treated bed nets.[35] The use of broad-spectrum insecticides, such as DDT, are also now known to have had a number of adverse effects on ecosystems across the globe. Their widespread use kills not only the targeted insect pests but also an array of beneficial predators and competitors.[36]

All of these observations by natural scientists and public health scholars aptly link human activities to higher incidence of infectious disease. But while humans make their own history, they do not make it as they please.[37] Our actions are shaped by the societies in which we live. Taking this into account, a common historical thread sews together the most pointed drivers of contemporary infectious disease outbreaks and the inadequate responses to them. The increasing toll of anthropogenic environmental change and our dwindling ability to prevent and cure infectious diseases have occurred amid the rising power and profits of Wall Street. Indeed, contemporary finance has shaped the lives of nearly every person and pathogen on the planet.

## THE SCOURGE OF FINANCE

Over the past two decades, discussions of finance and financialization have become more prevalent in mainstream news. In May 2023, Gillian Tett, chief of the editorial board of the *Financial Times*, penned a column titled "We Should All Be Worried about the 'Financialisation' of Our World."[38] Christine Emba, a columnist at the *Washington Post*, expressed a similar concern in April 2016 when she kicked off a series of op-eds with a piece titled "Has Our Economy Become Too Financialized?"[39] And in June 2014, Steve Denning, a senior contributor to *Forbes*, took a less pondering tone in an op-ed titled "Why Financialization Has Run Amok."[40] These are just a few of the many articles that have discussed financialization in the popular press. But despite appearing in well-known venues, the term *financialization* is unfamiliar to most.

Among both the media columnists and the academics writing about financialization, the term is often used to refer to the late-twentieth- and early-twenty-first-century trend in which finance has come to occupy a growing role in the economy. However, many scholars see financialization as a recurring process within global capitalism. In the words of economic sociologist Greta Krippner, financialization is "a pattern of accumulation in which profits accrue primarily through financial channels rather than through trade and commodity production."[41] As political economist Gerald Epstein further elaborates, during eras of greater financialization, "financial motives, financial markets, financial actors, and financial institutions" tend to play an outsized role in the organization of society.[42] In this view, the term *financialization* refers more broadly to periods of time when society and the world are increasingly shaped by and for finance.

Scholars of financialization often trace its contemporary origins to a series of policy shifts that occurred in the 1970s. After World War II, the United States held close to 70% of the world's gold reserves, making the US dollar by default the global currency. Global economic stability thus essentially depended on the United States carefully managing the circulation of a seemingly finite amount of its currency. On a global level, this process entailed efforts by the United States to track and account for the quantity and movement of US dollars outside its borders. Within the United States, banking regulation relegated responsibility for overseeing the circulation and distribution of money to the financial sector.[43] However, in the early 1970s the United States abandoned the gold standard, adopted a floating exchange rate, and more broadly began to liberalize its financial markets.[44] Other nations gradually followed or were forced to follow suit, putting into place floating exchange rates and regulations that encouraged capital mobility. In turn, an increasing amount of finance began to flow across the globe and capital became seemingly abundant.[45] Less and less restrained, finance and its benefactors began to remake the global economy.

Indeed, the 1970s saw the beginning of an era of vast financial innovation during which bankers and financiers created an array of new financial products and expanded the use of existing ones.[46] Perhaps most notable was the growing use of novel forms of financial securities. Historically, governments and corporations issued securities in the form of bonds and later stocks to fund major infrastructural projects and to acquire startup capital. However, the liberalization of financial markets allowed banks and financiers to create an array of new securities, many of which were composed of pooled income- or interest-generating assets. Over the past fifty years, banks and financiers have securitized everything from home mortgages to road tolls and music royalties, creating financial assets that today can bought and sold in the marketplace.[47] Such securitization has also been accompanied by increases in the trade of financial derivatives. Financial derivatives originated centuries ago as a means to protect against the risks of long-distance trade. Buyers and sellers of commodities used derivatives to lock in the price and quality for the future delivery of a physical good such as coffee or oil. In the 1970s, the liberalization of financial markets proliferated the types and volume of derivatives being traded. New derivatives for things such as currencies and interest-bearing loans were developed. In addition, traditional commodity-based derivatives for goods became increasingly traded among parties with no vested interest in the physical commodities themselves. Speculators and commodity traders alike thus began to chase profits from buying and selling financial derivatives with little to no regard for trade in the physical good the derivative was meant to represent.[48]

Bankers and financiers used the liberalization of financial markets in tandem with the creation of new financial instruments to extend their reach in another way—through debt. In the postwar era, capital controls such as tariffs and taxes constrained investment outside countries' borders.[49] In addition, the perceived scarcity of capital meant that banks could issue credit only for roughly the amount of money they regularly brought in and had on their books through interest on loans and savings accounts.[50] However, the loosening of capital controls allowed banks to more easily invest and circulate idle capital across the globe.[51] In addition, securitization allowed banks to turn the loans and other interest-bearing time deposits they held on their balance sheets into things that could be traded and offloaded on Wall Street. Through the creation of mortgage-backed securities and other financial products, banks thus gained more immediate access to additional capital that they could then recirculate and profit from.[52] As increasing amounts of financing—from credit card debt to sovereign loans and microfinancing—became available, people and places across the globe took on more debt. Nation-states and everyday people alike increasingly became shaped by their ability to acquire and pay off debt and to become what economic geographer Sarah Hall and others have called "financial subjects."[53]

Beyond financial innovation and debt, the growing amount of capital flowing across the globe and into financial markets also altered how corporations were valued and, in turn, how they operated. Instead of being valued for what they made or provided for society, corporations increasingly became valued for the returns they gleaned for shareholders. As sociologist Madeleine Fairbairn writes, starting in the 1970s, shareholder value "became the new shibboleth of corporate management."[54] As this shift occurred, corporations sought out new and enhanced cost-cutting measures and alternative sources of revenue. Cost-cutting measures took a wide array of forms, from eliminating underused assets and outsourcing, to offshoring and cutting worker pay and benefits. Attempts to find new sources of revenue frequently entwined nonfinancial firms with Wall Street. For example, some companies sought to boost sales by offering consumers loans, such as "buy now, pay later" installment plans, that were later made into financial products and then bought and sold in the secondary marketplace. This search for ever-increasing amounts of shareholder value resulted in the prioritization of profits for investors above all else.

We are often told that the proliferation of finance and the broader financialization of life provides us with greater opportunities to benefit from the profits of Wall Street. In the United States and elsewhere, the contemporary era of financialization has indeed created more investors and expanded the supply of credit. Since the 1970s, the percentage of households in the United States that report owning stocks has increased from around 25% to over 60%.[55] In addition, the International Monetary Fund (IMF) calculates that global debt, which includes debt held by public entities, households, and nonfinancial corporations, has more than doubled from around US$115 trillion to US$235 trillion.[56] The opportunity to both hold a diversified investment portfolio composed of stocks and other financial instruments and have access to increasing amounts of credit has thus created a number of beneficiaries. Perhaps most notably, the number of millionaires and billionaires in the world has substantially increased over the past forty years.[57]

However, the adverse effects of financialization have also been well documented. As academics and other analysts have demonstrated, the increasing dominance of finance in society has resulted in repeated economic crises and created an array of social problems.[58] Inequality in the United States has increased to levels not seen since the Gilded Age. In the words of James Lardner, "Financialization has been the elephant in the room of economic policy debate—a huge contributing factor to the skyrocketing incomes of a few and the nonliving wages of many, and a force that helps explain our neglected infrastructure, underfunded schools, outlandishly expensive colleges and the phenomenon of graduates impoverished by the high-interest loans that banks thrive on."[59] Indeed, financialization has resulted in a wide-scale redistribution of wealth. Financiers and their accomplices who make up the world's wealthiest 1% have grown their investment portfolios by draining the pockets of the average person and the coffers of public entities. At the same time, the power of finance has encroached on and taken over nearly every aspect of our lives. As Gautam Mukunda explains, the financial sector "has power, and real power doesn't just give you the ability to pay or force people to do what is in your interest—it reshapes the way people think so that they want to do what is in your interest."[60] In other words, the power of finance has become hegemonic, shaping our interactions in ways that strengthen the grip of Wall Street on society, often at the expense of our everyday lives.

But finance has not only shaped people. In its attempt to make everything into an asset that can be bought and sold on Wall Street, finance has remade the planet. In the simplest way, it has made nature into an asset that can be bought and sold in the financial marketplace. Moreover, even the development of seemingly non-nature-based assets has altered our planetary ecologies and created an array of problems. Indeed, contemporary finance can be credited with the creation of both landscapes that increasingly make us sick and the purported solutions that fail to make us well.

## FINANCIALIZATION AND INFECTIOUS DISEASE

Financialization has clearly generated an array of social ills. However, few have ventured as far as to say finance is pathogenic. Labeling finance as such makes sense, however, once we understand its hand in creating the conditions in which infectious diseases can emerge, spread, and even thrive. Gaining this understanding entails moving beyond natural scientists' observations that anthropogenic change is driving infectious disease to recognizing how broader political economic decisions shape and reshape society, the environment, and the relationships between them. More specifically, it requires understanding how the proliferation of financial securities, the mounting tolls of debt, and the prioritization of shareholder value above all else can create ecologies more prone to disease outbreaks and inhibit our ability to respond to them.

Political ecologists and scholars adopting what is called a structural one-health approach provide a productive starting point to make these connections. In a broad sense, political ecologists recognize how the social and material worlds in which we live are intrinsically entwined and dialectically produced to form what is commonly referred to as nature.[61] In examinations of health and disease, political ecologists have used such observations to identify the political economic factors shaping the ecologies both around and within us. More specifically, they have shown how capitalist-driven landscape change—from monoculture crop production and intensive resource extraction to the creation of urban shantytowns—has resulted in the formation of ecological niches known to enhance infectious disease.[62] Furthermore, political ecologists have demonstrated that the political economic processes shaping our material surroundings also influence how we respond to infectious diseases, and thus our bodies, our immune systems, and our ability to survive.[63] Structural one-health scholars extend these observations. Seeking to more directly address the inadequacies of natural scientists' and public health scholars' attempts to account for the social determinants of health, scholars utilizing such an approach suggest that conventional "one health" studies often fail to account for the broader political economic factors that shape human and nonhuman interactions that lead to infectious disease outbreaks.[64] To overcome this deficit, Rob Wallace and his colleagues suggest that scholars of infectious disease should more closely trace disease outbreaks "along circuits of capital."[65] A structural one-health approach can thus help scholars not only better recognize the structural dynamics contributing to infectious disease outbreaks, but also refocus studies seeking to understand the causes of such outbreaks beyond the sites in which they occur.

In this book, we build on the work of political ecologists and structural one-health scholars. In doing so, we seek not only to trace infectious diseases along circuits of capital but to understand their contemporary proliferation as a result of the broader financialization of society. We recognize that finance has long underwritten the landscape changes impacting infectious disease outbreaks and the societal responses to them. However, the contemporary increase in infectious disease suggests that the way this underwriting process unfolds has changed. To understand this change, we draw on the work of scholars examining shifts in the forms and flows of finance. These researchers demonstrate that financialization has impacted nearly every facet of life, and we use their insights as to how the increasing dominance of finance has shaped the landscapes in which we live, our health, and the science meant to solve the problems associated with infectious disease. As scholars of financialization have shown, making everything from mortgage debt to farmland into tradable securities with shares that can be bought and sold on Wall Street has altered urban, suburban, and rural landscapes, as has increasing the amount of credit available to people, private corporations, and public entities.[66] In addition, the prioritization of shareholder value in the provision of health care and the making of science has affected both the availability and the types of treatments used to prevent and cure human illnesses.[67] Bringing such observations into our own work, we identify the ways financialization has shaped the contemporary spread, frequency, and intensity of infectious diseases across the globe.

Recognizing the impact of financialization on landscape change, health care, and science allows us not only to understand the contemporary increases in infectious disease incidence but also to see the potential avenues toward a less disease-burdened world. Occurring within and as part of the broader financialization of society, the contemporary increases in infectious disease rates can be seen as the result of a distinct historical form of global capitalism. As such, the problems we currently face from infectious diseases should be seen as neither intrinsic nor permanent. As scholars of financialization demonstrate, the seemingly inescapable power that finance holds in and over our lives is granted and upheld by public institutions.[68] The outsize role of finance in society, and its impact on infectious disease, can thus be changed. Taking this prospect into account, we seek to demonstrate how finance can and has historically been used in ways that temper the outbreaks and effects of infectious disease across the globe.

## STILTGRASS, MORTGAGE BROKERS, AND COFFEE

As scholars of global processes, we have long been aware of the negative impact of financialization on society. In addition, we have both seen firsthand the adverse effects of infectious diseases on people and communities. However, the links between financialization and infectious disease have not always been at the forefront of our minds. It was not until one of us—Brent—decided to sell a house with a yard filled with ticks that the connections between finance and infectious disease began to unveil themselves. The existing expertise of the other of us—Kelly—on mosquitoes and malaria then helped to solidify the global nature of the problem.

In the early 2010s, Brent attempted to clear a wooded hillside in front of his Williamsburg, Virginia, home of Japanese stiltgrass, an invasive weed with no local predators. The weed was easy to identify and, with shallow roots, could be pulled up without much effort. However, amid the stiltgrass lived another species that was harder to see. After many days of pulling out the invasive species, Brent would find a tick or two on his clothing or attached to his body. At the time, he thought nothing of it. But a chat about stiltgrass and ticks with his colleague biologist Matthias Leu raised his concern. Matthias had started a preliminary study examining ticks and tick-borne disease and noted that he seemed to find fewer ticks in stiltgrass patches. He hypothesized that this was because deer—the primary local hosts for ticks—do not eat stiltgrass and thus spend less time grazing on it. Mentioning the presence of tick-borne diseases in the area, Matthias also warned Brent to do everything he could to keep the ticks from biting him. The following spring, Brent started using insect repellents on his clothing and tucking his pants into his socks when he went out to do yardwork. He also started to ponder the possible social drivers that created the landscape of his neighborhood as he tried to rid his yard of the still-pervasive stiltgrass.

Brent's curiosity eventually led him to wonder if a sociologist had ever studied ticks and tick-borne disease. After some searching, he found mention of Lyme disease in studies of what sociologist Phil Brown and his colleagues have called "contested illnesses."[69] In addition, he stumbled across a network of local support groups in northern Virginia for people who had struggled to find a proper diagnosis or treatment for the disease.

Seeking to understand the social dynamics surrounding tick-borne disease, Brent then began to travel to northern Virginia to interview public health officials, doctors, and local support group members about their experiences with Lyme disease. While most of his early interviews focused on the difficulties people faced when suffering from Lyme, many of the people he talked to brought up the ticks located right outside their homes. After such conversations, Brent could not help but think about the ticks he repeatedly pulled off himself when ridding his own yard of stiltgrass.

But it was not until 2014, when Brent and his family thought about moving, that he began to connect the broader shifts in finance to the ticks in both his yard and those of the people he spoke to in northern Virginia. When looking up the land records for his own home, he came across a series of aerial photographs of his neighborhood and lot dating back to 1937. For much of the twentieth century, the area was wooded. Segments of the land appeared to have been cut for logging purposes, but the forest was always replanted. However, in 1976, the first roads and houses were built in his neighborhood. The residential development filled in slowly at first, with a few dozen houses, including Brent's, being built. But between 1980 and 1986 the neighborhood rapidly expanded. Reflecting on his trips to northern Virginia, Brent realized that his own neighborhood was not that different from the neighborhoods of several of the people he had spoken with about Lyme disease. Indeed, the aerial imagery of Brent's neighborhood painted a picture of the ideal tick environment and what he would soon come to recognize as the stereotypical 1980s large-lot suburb in Virginia.

In his search for a new house, Brent also began to work with a broker to get preapproval for a mortgage loan. When inquiring about a loan, he was awed by the dizzying array of options available to him beyond the traditional thirty-year mortgage. While he knew a lot about the Wall Street takeover of mortgage markets, and though his existing mortgage had been bought and sold several times over, he had never thought about how he could convert his own home's equity into an asset that could then be invested in another securitized home loan. He also wondered how any of this was even possible. The financial crisis of 2008 in part stemmed from the use of shady mortgage-backed securities. The passage of the Dodd-Frank Act in 2010 was supposed to more closely regulate mortgage

markets. After seeking out a loan to buy a new house, Brent started to connect more of the dots in his research. Like the people who built the house
he wanted to sell and the people he spoke with in northern Virginia, Brent
was contributing to the financialization of residential landscapes and in
turn shaping the environment in which he lived.

In 2016, Brent moved to a house with a smaller yard, where, to this
day, he has never found a tick. His move was not inspired by a concern
about ticks. However, through his continued research in northern Virginia,
he began to realize that he had not fully escaped the scourge of insect-
borne diseases. In his research, an interview with a former public health
official turned at one point to the threat posed by mosquitoes. The former
official expressed concern about Lyme and other tick-borne diseases, but
having worked in Central and South America prior to living in northern
Virginia, he noted that he had a greater concern about the proliferation of
mosquitoes in Virginia that can carry chikungunya, dengue, and yellow
fever. Upon returning home, Brent identified the mosquitoes biting him
in his new backyard as the species *Aedes albopictus,* potential vectors for
each of those diseases. Shortly thereafter he stumbled across an article
reporting that the *Aedes aegypti,* another vector for the three diseases, had
been found in Washington, DC, and was able to overwinter in subterra-
nean structures such as subways and parking garages.[70] Over the following months, every time Brent went outside to tend his garden, he could
not help but notice a plethora of potential *Aedes* mosquito egg-laying sites.

As he began to think too much about how he could remove the small
puddles of water in his gutter downspouts, flowerpots, and children's toys
left out in his backyard, another disease carried by *Aedes* mosquitoes
arrived in the Americas: the Zika virus. While the Zika virus never established itself in Virginia's mosquitoes, Brent began to take notice of its
movement across Brazil. The disease was initially isolated to coastal urban
areas where the *Aedes aegypti* and dengue had long been known to wreak
havoc. However, incidence rates of Zika in Brazil's more rural and agricultural interior slowly began to overtake those in the country's urban
metropolises. Familiar with the financialization of Brazilian farmland,
Brent began to wonder if the same processes driving tick-borne disease in
Virginia were in some way driving mosquito-borne disease in central
Brazil. After noticing that the expansion of agricultural land in Brazil over

the past forty years took a strikingly similar trajectory as the spread of Zika, Brent expanded his research to include *Aedes aegypti* mosquitoes and the Zika and dengue viruses.

As Brent was clearing stiltgrass and thinking about the insects and the diseases they could carry in his backyard, Kelly was already examining the structural and ecological characteristics driving increases in infectious diseases. For years, Kelly had been researching how growth in agricultural exports from poorer countries led to rising rates of deforestation and bio-diversity loss. Through her studies, she recognized that the expanded production of a handful of agricultural commodities was responsible for a disproportionate amount of land transformation globally. While initially identifying this trend in the interwoven production of soy and beef, her studies drew her attention toward another agricultural commodity: coffee. Kelly did not drink coffee at the time, but watching her friends and colleagues regularly consume it helped her realize that coffee was a near-perfect example of the processes she was seeing in her studies. Coffee is one of the most traded and consumed commodities in the world, yet it is purely a luxury crop and fundamentally cannot be grown in the countries where it is consumed most conspicuously. The clearing of forests for the production of coffee was thus often out of the sight and minds of even the most socially and environmentally conscious coffee drinkers.

While contemplating the contradictions embodied in the coffee-drinking activities of her friends and colleagues, Kelly began to wonder if the increases in agricultural exports from poor countries adversely impacted not only local environments but also human health. Scholars had long demonstrated that poorer countries exporting primary commodities—be it raw materials or agricultural goods—often became mired in poverty.[71] In addition, the connections between poverty and ill-health were well established.[72] Yet, after decades of being told by free market economists and their echo chambers in the mainstream media that poorer places will most quickly develop if they exercise their so-called competitive advantages, the fact that increases in primary commodity exports did not necessarily result in a rising tide that lifts all boats mostly went unnoticed by the everyday person. So too did the increasing incidence rates of a number of infectious diseases that once again began to plague the poorest people in the world. Such ignorance, or perhaps indifference, was even present in Kelly's daily interactions. When she mentioned her interest in examining how macroeconomic policies impacted rates of tuberculosis and malaria, people all too frequently responded by asking if such diseases were even around much anymore. Over the following years, Kelly dedicated a fair amount of time demonstrating how macroeconomic policy change resulted in poverty, environmental degradation, and the spread infectious diseases. As she thinks back on the responses she received when initially talking about her research, Paul Farmer's remarks written decades before ring all too true: neglected diseases are forgotten once they stop bothering the wealthy.

In 2013, Kelly's interests in coffee and so-called forgotten diseases converged. Seeking to further ground the observations of her studies in the lives of everyday people, she began making annual research trips to the Bududa District of Uganda. Upon arriving in Bududa for the first time, she found one of the young children in the house where she stayed sick with a high fever and body aches, unable to get up from the couch. Kelly asked the female head of the household, the child's grandmother, what was wrong with the girl. The grandmother responded without much concern that the child had malaria. When Kelly probed and asked about needing to take her to a clinic or health facility, the grandmother replied that she would do her best to "buy her some treatment tomorrow." Kelly at first was shocked and somewhat concerned that the family, which she knew to be relatively wealthy and well respected in the community, would be so blasé about their granddaughter having a life-threatening disease. But after spending many cumulative months and years in Bududa, Kelly came to understand that malaria and mosquitoes are part of the fabric of everyday life in rural Uganda and the formal health care sector does not supply people's needs for malaria care and treatments.

Just as malaria is a part of daily life in Bududa, so too is coffee. Traveling through the region for the first time, Kelly saw coffee everywhere. But in contrast to the roasted and brewed coffee conspicuously consumed by Kelly's friends and colleagues in the United States, the raw bean was the most ubiquitous form of coffee in Bududa. Thousands of coffee plants lined the steep hillsides that had been taken over by plantations, and nearly everyone grew it in their gardens. During the harvest season, millions and millions of hand-picked beans covered the ground of community soccer fields and trading centers, laid out to dry on large sheets and tarps anytime the sun peaked through the clouds. As Kelly began to talk with local community members, in particular coffee farmers, the stories she heard aligned nearly perfectly with the observations she had made over the previous decade in examining macroeconomic shifts, global trade flows, and poverty. While the flows of global finance entering and the coffee exports leaving Uganda had increased over the past forty years, the Bududa coffee farmers had seen their incomes become more and more precarious, they told Kelly. But a somewhat unexpected story also appeared in Kelly's conversations. When talking about malaria, the farmers often noted that the mosquitoes that carry the disease preferred to live in their coffee gardens. Kelly then began to see how changes in coffee farming and a lack of proper access to health care in Bududa reshaped local community members' vulnerabilities to the disease.

While our research interests ran somewhat parallel for years, they converged in early 2021. Brent was teaching a class on the social determinants of infectious diseases in which he assigned Kelly's work. He realized that the same social processes he had observed driving the spread and greater incidence of Lyme disease in Virginia and the dengue and Zika viruses in Brazil likely underpinned the increasing rates of malaria in Uganda. Never having formally met Kelly, Brent sent her an email to see what she thought. After a discussion via Zoom, it became apparent not only that we were examining similar things in our research and teaching, but that the causal mechanisms underpinning the increases in and inadequate responses to vector-borne disease in each place were products of a broader global process: financialization.

## THE PATHOGENS OF FINANCE

On the surface, the emergence of Lyme disease in northern Virginia, the outbreaks of the Zika and dengue viruses in Brazil, and the endemic existence of malaria in the Bududa District of Uganda could not be more disparate. The counties in Virginia where Brent began his study of Lyme disease house some of the wealthiest people in the world on almost hidden, large-lot, suburban estates. The places with the highest incidence rates of Zika and dengue in central Brazil are occupied by the owners and workers of large-scale agricultural operations and surrounded by never-ending fields of soy. And the Bududa district of Uganda that Kelly frequents is home to some of the poorest people on the planet, many of whom live by maintaining small agricultural plots in which they grow coffee for cash and food for subsistence. While the social and material landscapes of these places could not look more different, the infectious diseases afflicting the people in each begin to draw them together.

Lyme disease, the Zika and dengue viruses, and malaria are all classified as vector-borne diseases. People are most often infected by each when a living organism—known as a vector—transmits the disease-causing pathogen between humans or between animals and humans. With Lyme, Zika, dengue, and malaria, the vector is a bloodsucking insect. Lyme disease is transmitted through the bite of the *Ixodes scapularis*, more commonly known as the blacklegged tick. The Zika and dengue viruses are transmitted through the bite of *Aedes* mosquitoes. And malaria is transmitted through the bite of *Anopheles* mosquitoes. The presence of each of these insect vectors has been altered by many of the characteristics that natural scientists attribute to the increase in infectious diseases more broadly. Climate change is expanding the range of these insect vectors and the frequency in outbreaks of the diseases they carry. In addition, landscape change—from deforestation to urbanization and suburbanization—has created more ideal environments for ticks and mosquitoes to both live and in turn transmit vector-borne diseases.

Recognizing the material characteristics of insect vectors, the diseases they carry, and the environments in which they can thrive tells only part of the story, however. The landscapes, livelihoods, and everyday interactions between people, pests, and pathogens are shaped and connected through the history of finance. For much of the twentieth century, states regulated finance in ways that fostered both the profit-making activities of private enterprise and the general well-being of their populations. From the strip malls of the United States to the strip mines of the Brazilian Amazon, states directed finance in ways that prioritized investments in infrastructure and technology not only to incorporate new places into circuits of capital accumulation but also to make these places free of vector-borne diseases. But in the 1970s, states shifted or were forced to shift from Keynesian to neoliberal forms of political economic organization. As this occurred, states began to prioritize profits for banks and shareholders over improving human livelihoods. In turn, the understanding that the prevention of vector-borne diseases required state intervention was replaced by the false promise that such prevention would occur at the behest of an invisible hand. This shift resulted in the creation not only of the ecological conditions in which vector-borne diseases are more likely to emerge and spread, but also of the social conditions that make the prevention, treatment, and eradication of vector-borne diseases more difficult. Understanding the recent increases in vector-borne diseases thus requires knowing how finance has shaped and reshaped societies, ecologies, parasites, and pathogens.

To pursue this inquiry, we make what Philip McMichael calls an "incorporated comparison."[73] Incorporated comparison is a method utilized to understand "social phenomena as differentiated outcomes or moments of an historically integrated process."[74] Incorporated comparisons can be made across time or space, or both. Cross-time comparisons look at "temporally differentiated instances" of a world historical process. Cross-space incorporated comparisons are used to examine the ways different places are linked as parts of a global social process.[75] The social phenomena we examine in making our incorporated comparison are the contemporary increases in and the responses to vector-borne diseases. To better understand these social phenomena, we make a cross-time and a cross-space incorporated comparison. In our cross-time comparison, we examine how global finance changed from the Keynesian and developmental state era to the contemporary liberal economic era and, in turn, shaped the landscapes of and societal responses to vector-borne disease. In our cross-space comparison, we examine how vector-borne disease outbreaks and the local responses to them in the United States, Brazil, and Uganda are shaped by and linked through global finance.

To make this incorporated comparison, we bring together historical, statistical, interview, and ethnographic data that we have collected over the past decade as part of our respective examinations of the social drivers and responses to vector-borne diseases throughout the world. Our observations on the changing forms of global finance and the contemporary turn toward financializaton are derived from our analyses of documents from government entities, corporations, and international agencies and an array of secondary sources. To recognize the impact of global finance on vector-borne disease, we ground this data in place through our interview and ethnographic data, analyses of local periodicals, and publicly available statistics on landscape change, infrastructure, demographic shifts, and disease incidence rates. In total, we collectively interviewed more than a hundred people and analyzed thousands of pages of documents from government agencies, colonial archives, international lending agencies, nonprofit organizations, and other sources. We utilize this data and our incorporated comparison to tell a story of financialization and vector-borne disease in three parts.

First, we explore the mechanisms driving the contemporary incidence increase and spread of vector-borne diseases. We argue that the liberal economic transformation of the global financial system provided the monetary foundations for massively altering the planetary landscape in ways that made vector-borne diseases more likely to spread and emerge. To make this argument, we trace and link the histories of landscape change and its impact on vector-borne diseases in the three places of our research from the colonial era to today. In the United States, we demonstrate how changing mortgage regulations and the creation of mortgage-backed securities resulted in the growth of large-lot suburban areas in Virginia and in turn the emergence of Lyme disease (chapter 1). In Brazil, we show how changes in international lending deepened inequality in urban areas and redirected global financial flows toward the country's interior, creating more ideal spaces in both Brazil's coastal cities and agricultural municipalities for mosquito-borne diseases such as dengue and Zika to thrive (chapter 2). And in Uganda, we illustrate how conditional loans, structural adjustment programs, and the increasing speculative trade of coffee derivatives contributed to increases in monocrop coffee production, deforestation, and increasing rates of malaria in the Bududa region (chapter 3).

Second, we examine the responses to the threats of Lyme, dengue, Zika, and malaria. In doing so, we expand our observations on financial shifts to examine their impact on science, medicine, and public health. We argue that the expansion of finance into global public health created a system more focused on securing shareholder value in the insurance, biotech, and pharmaceutical industries than on treating human illness, and, in turn, increasingly placed blame for the spread of vector-borne diseases on individual human and insect bodies. In the United States, we show how changes in regulations governing pension plans created a health care system built on increasing the shareholder value of medical insurance companies, making it more difficult for some people suffering from Lyme disease to receive adequate treatment and in turn placing responsibility for illnesses on the infected (chapter 4). In the case of Brazil, we demonstrate how shifts in the funding of both public health and the creation of scientific knowledge fueled by venture capital resulted in a greater focus on commoditizing insect bodies as the solution to dengue, Zika, and other mosquito-borne diseases (chapter 5). And, as for Uganda, we illustrate how the drive to simultaneously recuperate interest on foreign loans and generate shareholder value for the pharmaceutical industry changed the funding for health care and shaped the development and accessibility of antimalaria medications, making medical services unaffordable for many and turning the prevention and treatment of malaria into an individual responsibility (chapter 6).

Third, we draw on our observations about financialization and its consequences for landscape change, vector-borne disease, and our societal responses to the latter to envision a way to create a more ecologically sustainable and just planet on which vector-borne disease is both less likely to occur and less likely to impose severe harm (chapter 7). In exploring this prospect, we move beyond solutions that focus on individual human behavior as the best way to prevent and cure vector-borne disease. In addition, we attempt to resist the temptation to idealize the mid-twentieth-century era when vector-borne disease abated across the globe. Seeking to address head on the ways financialization has driven increases in vector-borne disease, we explore how the democratization of finance could result in the creation of new political ecologies less prone to Lyme disease, the dengue and Zika viruses, and malaria.

While our focus in this book is on vector-borne diseases, the processes of financialization we identify as underpinning their increases and as hampering attempts to prevent and treat them are applicable to infectious diseases more broadly. As blood-borne pathogens spread through bites of insects, the means through which Lyme, dengue, Zika, and malaria spill over into and are transmitted between human populations differs from airborne infectious diseases such as COVID-19, SARS, or H1N1 and from water-borne pathogens such as E. coli or cholera. However, from deforestation to degrading infrastructure, the ecological changes known to be driving increases in vector-borne diseases have also been demonstrated to be boosting air- and water-borne infectious diseases.[76] In addition, global shifts in the provision of health care and preventative public health measures have impacted the ability of communities and individuals to find solutions to health problems writ large. Taking this fact into account, we add an insect to the equation in our examinations of the ecologies of the vector-borne diseases, but the financial processes we link to their increased occurrence and the difficulties in preventing and treating them plague society's ability to address the problems associated with infectious disease more broadly. With finance influencing the ways in which we shape and interact with nature and with one another, the contemporary rise in infectious disease outbreaks and our inability to resolve them can plausibly be attributed to and labeled the pathogens of finance.

# 1 A Northern Invasion

In June 2015, Brent sat down with Michael Farris to talk about ticks and Lyme disease in Virginia. Farris's past made him an unlikely source of knowledge about ticks and tick-borne disease. As a constitutional lawyer and Christian conservative, Farris is best known as an advocate of religious freedom and homeschooling. For decades he defended both in court. He also founded Patrick Henry College in 1998 as an institute of higher education designed for Christian homeschooled students. In the political arena, Farris ran unsuccessfully for lieutenant governor of Virginia in 1993. After his defeat, he established a political action committee called the Madison Project to counter Emily's List and support pro-life economic conservatives dedicated to limited government and so-called free markets.[1] In 2010, such efforts placed him in the good graces of the newly elected Republican governor of Virginia, Robert McDonnell, whom he convinced to fund a statewide commission examining Lyme disease.

According to Farris, his wife and several of his children suffered from Lyme disease and struggled to get the appropriate treatment. Believing that the Centers for Disease Control and Prevention's (CDC) guidelines for the diagnosis and treatment of the disease were flawed, Farris wanted to change state and federal regulations to provide doctors with more leeway to use alternative treatments without fear of prosecution. Farris's work and positions on Lyme disease were much maligned by liberal media outlets such as *Mother Jones, Slate,* and the *New Yorker.*[2] Reporters frequently contextualized Farris's critiques of the CDC guidelines as part and parcel of his creationist beliefs, portraying him as an anti-science crusader. In Brent's conversation with Farris, he did make a number of unconventional and unproven suggestions about the transmission of Lyme disease, but he also had a deep understanding of the mainstream scientific literature on it. Indeed, his questioning of the CDC guidelines seemed equivalent to that of many activists engaged in struggles around un- or misdiagnosed diseases, or what sociologist Phil Brown and his colleagues call contested illnesses.[3] However, Farris was also infamously known for using the problematic discourse of Confederate apologists when describing the origins of Lyme disease in Virginia. Echoing in conversation what he had previously been quoted as stating in the liberal print media, Farris declared Lyme disease to be "a northern invasion" brought south by "those damn Yankee ticks."

In Farris's so-called "northern invasion," the perpetrators were a specific kind of tick—the *Ixodes scapularis,* more commonly referred to as the blacklegged tick. Blacklegged ticks have long been known to exist in the northeastern United States. One of the earliest recorded observations of the tick occurred near Cape Cod, Massachusetts, in the 1920s.[4] By 1945, the tick's presence had been sporadically reported in other states along the North Atlantic coast. Scholars examining tick distributions throughout the country, however, observed its largest concentrations in the gulf coast states.[5] It was only in the 1960s that more established concentrations of blacklegged ticks began to appear in the Northeast and in parts of the Upper Midwest. While scientists took note of the expansion of the tick's range, at the time it was not known to carry pathogens that had widespread effects on human populations and was thus not of high concern. It was not until 1970 that the blacklegged tick was found to carry *Babesia microti,* a parasite known to cause a malaria-like disease with fever and hemolysis.[6] Five years later, a mysterious disease also emerged in Connecticut that patients and doctors linked to ticks. It was eventually determined that the bite of a blacklegged tick could transmit the disease's bacterial agent, *Borrelia burgdorferi,* to humans, infecting them with what is commonly known as Lyme disease.[7]

The concentration of the black-legged tick in both the Northeast and along the gulf coast would seem to cast some potential doubt on Farris's suggestion of its "northern invasion" of Virginia. In addition, while climate change has expanded the range of the blacklegged tick, natural scientists demonstrate that the focal point of Lyme disease has moved in a northwestern direction from its place of discovery.[8] Underpinned by the observation that the warming of cooler climates has resulted in more hospitable environments for blacklegged ticks, the idea that such ticks and Lyme disease spilled over into Virginia from the North seems even less likely. But what, then, brought blacklegged ticks and Lyme disease to Virginia, and where exactly did they come from?

For much of the twentieth century, blacklegged ticks were not known to exist in Virginia. Lyme disease began to appear in patients in the state in the mid-1980s.[9] In a 1995 study, 9% of Virginia counties reported the presence of a blacklegged tick, and 6% of counties reported established populations of the tick.[10] Over the following decades, the blacklegged tick became much more common in the state. By 2015, 53.7% of Virginia counties reported the presence of a blacklegged tick, and 32.1% had established populations.[11] As blacklegged tick populations rose, so did the incidence of Lyme disease. In more recent years, over a thousand cases of Lyme disease have been reported in Virginia annually. The counties with the highest caseloads have all been located in northern Virginia. And Loudoun County, where Farris lived, has regularly reported the most cases in the state.[12]

Farris's attribution of the increases in Lyme disease in Virginia to a "northern invasion" by the blacklegged tick are likely best seen as an extension of ill-fated revisionist histories of the Civil War. However, higher rates of Lyme disease and the spread of blacklegged ticks in the state are embedded in another type of invasion. Indigenous peoples long lived on the land that is now known as Virginia, but the invasion of the territory by European colonizers commenced a radical transformation of the landscape that would prove to reshape the presence of ticks and tick-borne diseases. Widespread deforestation starting in the seventeenth century likely lowered the threat posed by ticks and tick-borne disease in the state for a time, but in the mid- to late twentieth century the shifting forms and flows of housing finance in the United States and the expansion of the suburbs of Washington, DC, enabled the gradual transformation of north-ern Virginia's agricultural fields into residential landscapes. While post-war housing finance drove the expansion of the more densely populated urban areas such as Arlington and Alexandria, the late-twentieth-century shifts in housing finance drove the expansion of the large-lot estates of Fairfax, Loudoun, and Prince William counties. And in these sprawling sub- and exurban spaces, a new landscape was created in which ticks and tick-borne disease could thrive.

## THE EARLY HISTORY OF THE BLACKLEGGED TICK AND LYME DISEASE IN THE UNITED STATES

Blacklegged ticks and Lyme disease have long existed in the United States. Phylogeographic studies suggest the blacklegged tick originated in the southern United States over a half million years ago and expanded its range after the retreat of the Laurentide Ice Sheet 20,000 years ago. As it did, the tick is thought to have established a presence in most states east of the Mississippi River.[13] Studies have also shown that that the bacteria that causes Lyme disease has been present in North America for over 60,000 years.[14] However, natural scientists believe that human-induced environmental change dramatically altered the distribution of the black-legged tick and the diseases it carries. In particular, widespread deforesta-tion, the expansion of agriculture, and the hunting of white-tailed deer to near extinction in the 1800s and early 1900s are thought to have concen-trated the tick's presence to the few places where woodlands remained intact.[15]

The biophysical conditions necessary for blacklegged tick survival are not known to have changed dramatically over time. At the most basic level, the tick needs a relatively humid environment to live. It can exist in a dry area on its own for a limited period of time, but it will likely die if it does not return to a moist place within a day.[16] The blacklegged tick also depends on having an annual meal of blood at each of its three life stages in order to satiate its nutritional needs.[17] Forests serve as ideal environ-ments for blacklegged ticks. The shade prevents the ground from drying out quickly, and the frequent presence of dense leaf litter and brush provide a damp insulating cover for ticks to live in during dry- and cold-weather months.[18] Such forest conditions are also favorable to the presence of a number of viable tick food sources, or what ecologists call "hosts." Large and small mammals, such as deer and mice, as well as an array of reptiles, such as lizards and box turtles, often make their homes on the forest floor and sleep on or underneath the fallen leaves. Birds also frequent the forest floor in search of their own nutritional needs. As blacklegged ticks search for a bloodmeal, they can easily latch on to any of these animals as they rest or eat on or near the ground. To enhance their chances of encountering a food source, blacklegged ticks also engage in an activity that ecologists call questing. A blacklegged tick quests by climbing up the branch of a low-growing shrub, a stem of a plant, or a blade of grass. On reaching the end, it waits until it senses the carbon dioxide from the exhaling breath of a potential host. It then extends its front legs outward in a position that makes it ready to grasp on to a living being as it passes by and brushes against it.[19]

Blacklegged ticks can carry a variety of bacteria and viruses that cause disease in humans, but the ticks are rarely, if ever, born with harmful pathogens already present in their bodies.[20] Blacklegged ticks acquire the bacteria after feeding on an infected host. Such a tick then serves as a vector, carrying the bacteria from one host to another as it takes successive bloodmeals throughout its lifespan. Many people believe that white-tailed deer are the culprits behind the increased occurrence of Lyme disease in the United States. As large mammals, the deer do serve as the primary feeding and breeding grounds for blacklegged ticks. However, white-tailed deer are what ecologists call incompetent hosts and highly inefficient reservoirs for *B. burgdorferi*, meaning they are unlikely to host the pathogen even if they are bitten by a tick that carries it. Small vertebrate animals with poor hygiene, most notably the white-footed mouse, have been found to be the most "competent" hosts and efficient reservoirs, thus more frequently passing *B. burgdorferi* to ticks which then pass it on to humans.[21]

While the environmental and nutritional needs of blacklegged ticks are thought to have remained relatively constant for hundreds, if not thousands, of years, the places where they currently live have not. In particular, forested landscapes have been made and remade over the past four hundred years. Indigenous populations long shaped the landscape in distinct ways, but nearly all of the land east of the Mississippi was enshrouded by forests prior to the establishment of the first British colonies in the early 1600s.[22] These forested landscapes were filled with an array of wildlife that today are known hosts of the blacklegged tick. Indeed, environmental historians have noted that European explorers often wrote of the abundance of large and small mammals, such as white-tailed deer, elk, wolves, and gray squirrels, as well as a wide array of birds.[23] In addition, colonial explorers and settlers frequently noted the presence and annoyance of ticks.[24] Early colonizers and the Indigenous populations that preceded them thus likely experienced Lyme disease, but to what extent is largely unknown.

Ticks and Lyme disease may have existed in much of the eastern United States, but a number of factors likely limited the number of blacklegged ticks and thus the incidence of Lyme disease in precolonial times. First, precolonial forests are thought to have had higher levels of ecological diversity and species balance.[25] According to contemporary ecologists' insights, the presence of a greater abundance and diversity of wildlife may contribute to a "dilution effect."[26] Greater biodiversity in precolonial times could thus plausibly have increased the likelihood that blacklegged ticks fed on inefficient hosts and were thus less likely to carry Lyme disease. Second, both the hunting practices of Indigenous populations and the presence of natural predators likely prevented any single tick host from becoming overly concentrated. Third, historians also believe that Indigenous populations in the Americas regularly used fire to open up portions of the forest floor.[27] Drawing on the insights of contemporary conservation and wildlife scholars, researchers have deduced that controlled burns and forest fires reduced tick populations.[28]

While ecological diversity likely limited the presence of blacklegged ticks and Lyme disease in precolonial times, ecological destruction likely nearly eliminated the threat of both. In the early seventeenth century, the Indigenous practices that had shaped the landscape for centuries abruptly came to an end. European colonizers forcefully removed Indigenous populations from the land and turned forests into fields for cultivation and settlement. At the same time, timber became a vital building and energy supply. While much of the interior forests of the United States remained lightly touched, the coastal areas on the Atlantic and inland areas with access to water transport were significantly logged by 1850. The expansion of railroads at the turn of the twentieth century further linked untapped forest reserves in states on the gulf and Pacific coasts to the more populated urban areas in the Northeast where timber was already scarce.[29] However, by the 1920s, even these supplies of timber were being rapidly diminished.

The wildlife throughout the United States experienced similar levels of decline. European colonizers saw large mammals such as deer and elk as readily available food supplies. In addition, they saw certain natural predators as threats to both the livestock they brought to North America and the wild game they ate.[30] Early blame for the diminishing supply of desirable game animals that could be eaten was thus often placed on wolves and cougars. Indeed, some early European settlements placed bounties on wolves as early as 1630.[31] Local authorities later recognized that unchecked hunting diminished supplies of edible game. In 1646, local authorities in Portsmouth, Rhode Island, limited deer-hunting season to seven months of the year. By 1720, all of the colonies had followed suit by adopting similar closed hunting seasons.[32] Such checks did little to slow the mass slaughter of native wildlife in the Americas, however.[33] For example, historical estimates place the precolonial population of white-tailed deer at over 20 million. By 1800, the destruction of habitat and hunting by European colonizers is thought to have reduced the white-tailed deer population by 35% to 50%. By 1900, the population is thought to have plummeted to less than a million. Other wildlife species in the United States are believed to have suffered a similar fate.[34]

As settlers cleared the land and hunted large mammals to near extinction, the blacklegged tick became confined, it is thought, to the more remote woodland spaces that remained east of the Mississippi. Within the tick's historical range, these spaces likely mirrored those where white-tailed deer sought refuge.[35] In the northern states, this area included the less traveled spaces in the upper tips of Maine, New York, Michigan, Wisconsin, and Minnesota, as well as on near-shore islands off the north Atlantic coast. In the southern states, the area included both the rockier mountainous spaces and the wetland and bottomland forest areas of the gulf coast states. While the blacklegged ticks were able to survive in these places, the decreased presence of an inhabitable woodland environment likely constricted both its population and the number of incidents of Lyme disease across the United States as a whole.

However, in the isolated places where blacklegged ticks continued to exist, the presence of Lyme disease in the ticks and competent hosts may have actually increased. If a greater abundance and diversity of wildlife could result in the dilution effect, a lack of abundance and diversity could result in the opposite.[36] In other words, if spaces where blacklegged ticks and Lyme disease became confined had an abundance of competent hosts and a lack of incompetent hosts, questing ticks might be more likely to latch onto and draw a bloodmeal from a host that carries or could carry Lyme disease. Within this context, a tick's chances of both becoming a vector for the disease and passing it to a previously uninfected host would increase. Given the overhunting and near extinction of many large and medium-sized mammals known to be incompetent hosts, ticks left in wooded refugia were likely feeding more often on smaller mammals, such as white-footed mice, that are known to carry Lyme disease.[37] But as long as such refugia remained far enough away from more densely populated areas, incidents of Lyme disease in humans were likely low and may have gone unnoticed. These conditions would all begin to change with the more rapid expansion of the urban edge in the United States.

## THE SPILLOVER: FINANCE, LANDSCAPE CHANGE, AND THE MAKING OF TICK ENVIRONMENTS

The expansion of the urban edge in the United States is often linked to suburbanization. In the popular imagination, the suburbs are a postwar place of single-family homes, each surrounded by a lawn and a white picket fence. Such an image holds strong in the minds of many, but it is largely an ahistorical description of an ever-changing liminal space betwixt and between the urban and rural.[38] Suburbanization in the United States can be traced back to at least the mid-nineteenth century, and its form has varied greatly across space and time. While suburbanization was historically portrayed as an escape from the pests and pestilence of city living, certain forms have proven to enhance the likelihood of suffering from some diseases. In the suburbs linked to the twentieth-century

American Dream, the sedentary and car-dependent lifestyles of their inhabitants have been linked to health problems such as heart disease and obesity.[39] As forms of suburbanization in the United States changed toward the end of the twentieth century, however, new diseases emerged as potential threats to human health. In particular, as suburbia expanded outward and the size of many lots grew, landscapes were created that enhanced human exposure to an array of zoonotic and vector-borne diseases, including Lyme.

Economic elites were the first to seek out homes on the edges of cities. In the nineteenth century, homeownership began to replace yeoman farming as the central aspiration underpinning the purported American Dream. In turn, middle-class suburban neighborhoods began to emerge as early as the 1860s.[40] At the time, housing reformers saw densely populated urban tenements as "breeding grounds for disease and leftist politics alike."[41] Workers who owned a home and were encumbered by debt, it was thought, would be not only less likely to get sick but also less mobile and less likely to strike. As a result, homeownership and suburbanization were seen as ways to both create a nation of invested property owners and to pacify the workforce. The Great Depression temporarily derailed people's ability to achieve this American Dream, but the perceived benefits of homeownership for capitalists were not lost. Indeed, the federal government institutionalized widespread homeownership—and, by default, suburbanization—as a goal of the New Deal. As it did, homeownership became an instrument of statecraft through which the landscapes of postwar American suburbs were shaped by the regulated contours of finance.

New Deal housing and banking regulations propelled the expansion of the idealized White middle-class suburb of the American Dream by making credit in the form of mortgages more accessible to greater swaths of the population. One way this was done was through the creation of banks and lending institutions where none had previously existed. Prior to 1933, one-third of counties in the United States had no place to bank or obtain a mortgage. To resolve this problem, the federal government authorized savings and loan associations (S&Ls) to be federally chartered under the Home Owners Loan Act (HOLA) to ensure that people in all parts of the country had access to a local institution where they could bank and obtain or refinance a home loan. To ensure that S&Ls acted locally and for a greater number of people, they were allowed to offer mortgages only to consumers purchasing homes within fifty miles of the bank's home office, and ceilings were placed on loan amounts.[42] Through the Home Owners Loan Act, Congress also created the publicly owned and capitalized Home Owners Loan Corporation (HOLC) to purchase and refinance troubled mortgages on self-amortizing fifteen-year loans with 5% interest rates for up to 80% of a property's value. In 1934, the federal government further expanded mortgage credit availability through the National Housing Act and creation of the Federal Housing Administration (FHA). The FHA incentivized private lending for mortgages by backing defaulted loans.[43] Like HOLC loans, Federal Housing Administration loans were self-amortizing, were long term, could be extended with higher loan-to-value ratios, and had relatively low interest rates. As the number of both HOLC and FHA mortgages grew, such loan characteristics became industry standards and made home ownership more accessible to greater numbers of people. The extension of savings and loan associations and the relative standardization of mortgage terms and interest rates lowered the monthly payments on most home loans, making them more accessible to the average borrower and making default less likely.[44]

To further promote homeownership, S&Ls were systematically privileged over other types of banks though the Banking Acts of 1933 and 1935. In a broad sense, this legislation was passed to limit future risk in the financial sector by compartmentalizing the industry into three primary segments with separate responsibilities. Savings and loan associations were given primary responsibility for consumer savings accounts and lending. Commercial banks were confined to dealing with business accounts and loans. And the laws relegated securities and investment firms to making trades in the stock market.[45] A measure known as Regulation Q enabled S&Ls to more easily fulfill their role. Regulation Q placed ceilings on the interest rates of savings accounts offered by other types of banks. Savings and loans were thus able to offer higher interest rates and ideally attract most consumer savings deposits. Such deposits could then, in turn, be lent out as home loans.[46]

As these New Deal banking and housing regulations propelled mass suburbanization in the United States, they also constrained its form. Such policies made mortgages more widely available, but they were still designed around the premise that the world was credit short and capital starved. The prevailing idea behind policy makers' decisions was that there was only so much money to go around and that it was necessary to facilitate the rational distribution of scarce capital between different economic sectors and across the United States.[47] Within this context, incentivizing consumer savings in savings and loan associations by offering higher interest rates directed capital into the housing sector at the expense of other sectors. Not only did this strategy insulate the housing sector from full-fledged market competition and keep mortgage interest rates relatively low, but it also more equitably distributed credit and tempered the size of housing. In a context of scarce capital in which S&Ls prioritized local lending and offered mortgages on homes only up to a certain value, it was difficult for people to finance large homes on large lots. As a result, the mid-twentieth-century American suburbs were made up of modest-sized homes surrounded by a lawn yet still within earshot of a neighbor.[48]

While this suburban landscape would later be known to contribute to a host of chronic health problems, it did prove to be an escape from the infectious diseases present in the more densely urban areas of previous times.[49] Many mid-century suburban homes had full access to potable water and modern sewage and wastewater systems, decreasing both the number of places where infectious diseases and their hosts are born and the likelihood of human contact with such diseases. In addition, the increasing prevalence of single-family homes resulted in a greater spatial separation of people from one another, decreasing the probability of human-to-human transmission. Combined with the increase in vaccinations and the widespread use of DDT to kill disease-carrying insects, the suburban landscape proved to be relatively free of infectious diseases such as tuberculosis, malaria, and typhoid fever that plagued people in the past. And while the expanding urban edge did bring people closer to places where ticks and tick-borne disease likely existed, the form suburbanization took at this time largely left the parasites in their wooded refugia where they had retreated in previous centuries. Indeed, with many suburban neighborhoods composed of single-family homes stitched together by a semicontinuous mat of turfgrass, the environmental conditions for ticks and tick hosts were less than optimal.

Yet the seeds for an outbreak of Lyme disease were sewn as a credit crunch hit the banking industry in the mid-1960s. As inflationary pressures repeatedly drove the interest rates on US treasury bills above those offered on savings accounts, banks were unable to compete, and depositors began to withdraw their money to secure higher yields elsewhere.[50] In response, savings and loan associations and commercial banks began to explore new ways to bring in capital. Most notably, they developed new financial innovations—such as money market accounts and mortgage-backed securities—that allowed them to work with security and investment firms to gain access to more liquid capital from, and offload risk to, the secondary marketplace.[51] The emergence of such products in banking provided S&Ls and commercial banks with temporary relief from dwindling deposits. However, it often put them in competition with each other to attract depositors, undermining the stability of the banking sector while simultaneously allowing security and investment firms to rise to a place of dominance in the financial world. Indeed, dependent as they were on securities and investment firms to sell their products on the secondary market, savings and loan associations and commercial banks provided those firms with the opportunity to more directly enter and profit from the banking sector. However, securities and investment firms were not satisfied with merely being the recipient of funds from S&Ls and commercial banks. Seeking to cut out the middleman, securities and investment firms began to develop an array of their own financial instruments—from money market mutual funds to cash management accounts—to woo the higher interest–seeking clientele from traditional banks.[52]

With investment firms offering higher-yielding account options and luring in the everyday depositor, pressure to deregulate the banking industry began to mount. While large savings and loan associations and commercial banks had long been calling for the end of deposit rate ceilings and more freedom to offer a variety of investment and loan options, smaller banks that did not have the capital to develop and introduce new financial instruments pushed to maintain New Deal regulations that kept the banking industry separated. However, in the early 1980s, with passage of the Depository Institutions Deregulation and Monetary Control Act and the Garn–St. Germain Act, Congress acquiesced to the interests of large banks. The two acts eliminated interest rate ceilings on deposit accounts and allowed banks to offer loans at interest rates of their own choosing. In addition, the acts further blurred the lines between S&Ls and commercial banks, allowing each to diversify its holdings by offering products previously only sold by the other.[53] Neither of these changes, however, stopped the rise of investment and securities firms.

Indeed, the regulatory shifts in banking and finance during the 1970s and 1980s propelled investment and securities firms into becoming the dominant actors in the housing sector. Savings and loan associations had long been the primary originator of mortgages. Offering higher interest rates on savings accounts, S&Ls were set up to take in the majority of capital from the everyday depositor and turn such money into residential home mortgage loans. However, the changes in banking and finance allowed mortgage companies and brokers to originate home loans and sell them directly to securities and investment firms. Between 1970 and 1995, mortgage companies went from originating a quarter of all new mortgage debt to more than half, and S&Ls went from originating close to half of all new mortgage debt to less than a fifth.[54] As a result, increasing amounts of mortgage debt ended up in mortgage pools and trusts operated by securities and investment firms. Between 1970 and 1995, the proportion of mortgage dollars held in such pools and trusts increased from less than 1% of mortgage debt to nearly 40%.[55] Indirectly operating as service companies for securities and investment firms, mortgage companies thus essentially circumvented and obviated the need for savings and loan associations while facilitating the transfer of control over the housing industry to Wall Street.

As financial firms became the dominant actors in the housing sector, the purpose and ideology underpinning the ways the banking and mortgage industries operated changed from promoting the public good to rent seeking. In the early postwar era, federal regulations had been put in place to increase homeownership, which in turn served as a vehicle for the everyday family to accrue wealth. Such regulations limited competition between banks for deposits and allowed banks to offer mortgages with relatively low interest rates. From 1940 to 1965, interest rates on conventional home mortgages never rose above 6%, and home ownership rates steadily increased.[56] As interest rate ceilings were eliminated, however, banks began to offer higher returns on everyday deposit accounts.

To partially make up for their losses, banks raised interest rates on home loans. For most of the 1980s, mortgage interest rates thus hovered around 9%.[57]

High interest rates caused the characteristics of homeowners, their homes, and the broader residential landscape to change. In general, fewer people were able to purchase homes, and homeownership steadily declined until the mid-1990s, when interest rates decreased.[58] Homeownership did not decline among all segments of the population, however. Whereas homeownership increased across all income brackets in the twentieth century postwar era, from the late 1970s to the mid-1990s it declined among the lowest income deciles while remaining relatively constant or increasing among the top income deciles.[59] With homeownership confined to the better-off, the average square footage of new homes in the United States increased at one of the most rapid rates in history. From 1978 to 1995, the median new American home increased from 1,650 to 1,880 square feet and the mean increased from 1,750 to 2,050 square feet.[60] At the same time, more people began to seek out homes on larger lots. By the mid-1990s, lots larger than an acre accounted for 90% of the land used for new housing.[61] With more than 60 million people residing on such acreages, their parcels collectively made up more than 30% of the land in the contiguous forty-eight states.[62]

These sub- and exurban homes held many of the same characteristics that had made postwar suburban homes seemingly healthier places to live. In addition, the budding homes on sprawling residential acreages purportedly provided an escape from the mental and physical anguish of city living. Surrounded by trees, green grass, and an abundance of birds and deer, these sub- and exurbanites seemingly lived in nature. However, they did not build their homes in any sort of pristine environment. Instead, their houses were constructed along with and as part of the nature in which they were placed. The construction of such homes on the outer edges of urban centers frequently resulted in the creation of what ecologists refer to as fragmented forests. Forests classified as fragmented have large edge-to-interior forest ratios, smaller patch sizes, and a degree of isolation or distance from other forested areas.[63] The construction of large lot sub- and exurban houses created fragmented forests in a number of ways. Some homes were carved into already wooded landscapes and thus fragmented an existing forest. Others were built in open fields, landscaped with saplings and hedges for greater privacy, and thus contributed to the creation of a fragmented forest as such vegetation matured. In some places, such large-lot homes were also surrounded by generous easements that further separated homes and provided residents with even more access to green space. In turn, the fragmented forests on homeowners' lots gradually blended with those surrounding public property.

These emerging fragmented forests would prove desirable environments not only for wealthier homeowners seeking a larger slice of land with a bit more privacy and greater access to nature, but also for an array of ticks and tick hosts. As in most forests, the dense leaf litter in these residential forests provides a moist insulating cover for ticks and their smaller hosts.[64] However, forests with larger edges allow more sunlight to reach the forest floor, which promotes growth of lush green vegetation. Such vegetation frequently serves as an attractive food source for tick hosts, in particular the white-tailed deer. In addition, forests occurring in smaller patch sizes and in isolation from other forested areas often have lower species richness and thus fewer potential predators.[65] As a result, if a bacterium such as *B. burgdorferi* becomes established, it is more likely to spread in competent host populations. In the changing landscapes of the late-twentieth-century urban edge, the inhabitants of sub- and exurbia thus helped to create an ecosystem with an omnipresent bodily threat— bloodsucking ticks and the diseases they carry with them. The construction of Michael Farris's home and the fragmented forests that surround it in Loudoun County, and northern Virginia more broadly, can be seen as an illustrative example.

## THE NORTHERN INVASION: SPRAWL, TICKS, AND LYME DISEASE IN VIRGINIA

Northern Virginia has long been shaped by the politics and economics of the federal government. In this place of early settler colonialism in the United States, European colonizers began forcefully removing the Indigenous peoples as early as 1619. In the process, White settlers disturbed the land's ecological balance as they rapidly converted it to farmland for export-oriented cash crop agriculture. A century later, some of the northern territory of Virginia was granted to the federal government to build the nation's capital, only to be retroceded fifty years later. In the mid-nineteenth century, the region served as a buffer between Washington, DC, and the capital of the Confederacy in Richmond, making it the staging grounds for numerous battles during the American Civil War. And in the twentieth century, northern Virginia became home to some of the most powerful branches of the American national security apparatus, the Department of Defense and Central Intelligence Agency. All of these events in some way changed the landscape of northern Virginia. However, it was the purported small-government policies of Ronald Reagan that drove, and it was the changes to federal mortgage policies that made possible, the wave of suburban and exurban sprawl and, in turn, Michael Farris's so-called northern invasion.

Situated alongside the Potomac River and Chesapeake Bay, the forests of northern Virginia and of Loudoun County near Farris's home were likely some of the first to be felled by European colonizers. With these waterways serving as the primary means of commercial transport, the trees could be cut and easily shipped as lumber to wherever it was needed. Access to a major waterway eased the transport of agricultural commodities, as well, and made the area a prime location for plantations and the production of cash crops for export, in particular tobacco.[66] As a nutrient-intensive crop, tobacco in sustained production at this time required excess uncultivated land. Tobacco production usually depleted soils of essential nutrients in two to three years. Fields then had to be left fallow for the same period of time to regenerate. As a result, tobacco producers continually felled more trees in order to access soils with higher nutrient loads. While such deforestation was somewhat hampered in early colonial times by the labor-intensive nature of tobacco production, the use of enslaved Black laborers on plantations allowed tobacco producers to expand production and fueled the more rapid destruction of forested areas.[67] Within sixty miles of Farris's home, numerous tobacco plantations once existed, including George Washington's Mt. Vernon. The area had thus been significantly logged by 1850.[68]

As in other American colonies, the benefits of biodiversity in Virginia also went unnoticed by the European colonizers. They viewed wildlife predators such as bobcats and coyotes that hunted game animals as nuisance species and often killed them with little regard.[69] Larger game, such as deer, were overhunted for food, clothing, and the value of their hides in the transatlantic export market. While Virginia established a deer hunting season in 1699, a number of exemptions existed, and a general lack of enforcement made it difficult to prevent out-of-season hunting. Between 1600 and 1930, estimated deer populations in the state thus decreased from between 400,000 and 800,000 to approximately 25,000 animals.[70] During this time, most large game in and around Loudoun County were hunted to near extinction. Indeed, in the early twentieth century, deer were no longer thought to even exist in much of northern Virginia and returned only after being reintroduced to the region.[71]

The proximity of northern Virginia to the nation's capital long made it a potential place of urban spillover, but the area was largely neglected by the early founders of the United States in favor of development north of the Potomac.[72] One of the earliest expansions into the area occurred in 1864 when the federal government was looking for a place to bury soldiers who died during the Civil War. Already occupying Robert E. Lee's plantation, the federal government turned it into a burial ground that would later become Arlington National Cemetery. Over the following half century, US government activities in Virginia would largely be confined to placing the dead in their final resting place. However, as the war-waging activities of World War I and World War II created a demand for space to accommodate the living, northern Virginia was more fully incorporated as an appendage to Washington, DC.[73] During World War I, the large influx of people into the nation's capital as workers supporting the war effort outstripped the city's housing supply and sparked a population boom in Alexandria, Virginia. During the Great Depression, more workers flocked to the area as government-sponsored New Deal employees, and the population boom expanded outward into Arlington County. And during World War II, northern Virginia became more than just a bedroom community of Washington, DC, when the federal government took on a more physical presence by building the Pentagon in Arlington County in 1943.[74]

Much of the early southern sprawl from Washington, DC, resulted in densely populated communities housed in garden apartments, duplexes, and later, residential high-rises and townhomes.[75] Where new single-family homes were constructed, neighborhoods looked similar to other places in the United States. Some exceptions existed, but most new homes were modest in size and situated on a quarter acre or less of land.[76] But, by the 1950s, a slow trickle of political and economic elites began to purchase country homes in Virginia west and south of what was then outside the greater Washington, DC, metropolitan area. As historian Andrew Friedman notes, northern Virginia was often seen as a place that had "the beauty and charm of the countryside" but could "reflect in modern living the graciousness of the past."[77] In other words, Northern Virginia was a place with available acreages that reflected the state's White southern plantation past. Some of the best-known mid-twentieth-century postwar migrants to the area included Supreme Court justice Byron White from Colorado, Eleanor Dulles from New York, and perhaps most notably, then-senator John F. Kennedy of Massachusetts. Kennedy's Hickory Hill estate, which he sold to his brother Robert in 1956, sat on 5.6 acres that were surrounded by trees and foreshadowed the future sprawling residential homes of those settling the area.

The expansion of the federal government into northern Virginia would further contribute to suburban sprawl with the completion of the new Central Intelligence Agency (CIA) headquarters, or what is often referred to as Langley, in 1961. Unlike most government buildings in the nation's capital, Langley was constructed off the beaten path. It was built in a 750-acre wooded landscape in Fairfax County crossed by a creek and surrounded by dairy farms and a foxhunting forest. Prior to its development, the site had no access to major utilities. Indeed, before being applied to the CIA headquarters, *Langley* was "simply the name of a fork in the road," known largely only to local residents. However, while Allen Dulles—the director of the CIA during Langley's construction—tried to keep as much of the forest surrounding the agency's headquarters as possible in the government's hands, Langley's design was as much an attempt to construct nature as it was to maintain it. Indeed, while the headquarters was shrouded in trees that rendered it invisible from all roads, the 2.3 million–square foot building was situated on architecturally shaped slopes and hillsides and surrounded by both open grassy lawns and trails that sliced through the woods.[78]

An isolated building in the woods—even one as large as Langley—would likely not by itself create a landscape in such a way as to dramatically increase tick populations. However, just as Washington, DC, was the planning grounds for settler colonial military incursions in the United States that eliminated Indigenous people and transformed the landscape for White property owners in the 1800s, Langley served as the planning grounds for military incursions that inflicted harm on Black and Brown people abroad and its construction transformed the landscape to accommodate the largely White property owners who worked there. In the process, the construction of the CIA headquarters and subsequent migration of its employees to the area would plant the seeds for a more favorable and ever expanding tick environment in northern Virginia. In the 1950s and 1960s, the populations of Fairfax County, Loudoun County, and Prince William County all grew at rapid rates. The population in the three counties combined more than quadrupled from 142,316 in 1950 to 603,273 by 1970.[79] Although the area's early suburbs were relatively dense, concerns about rapid population growth and the erosion of northern Virginia's rural aesthetic led to opposition to residential zoning codes that would have set the maximum size of lots to one acre.[80] Taking advantage of the possibility of owning an acreage, many CIA officers built homes on the secluded lanes of Fairfax County. As Friedman notes, the location of the CIA headquarters helped "to preserve the character of the community, essentially providing CIA cover for the exclusionary large-lot zoning that made Langley such a desirable security location in the first place."[81]

In the 1980s, the presence of tick landscapes in northern Virginia increased as President Ronald Reagan purportedly sought to downsize the federal government. Seeing government as "the problem," Reagan formed the President's Private Sector Survey on Cost Control—also known as the Grace Commission—to identify ways the federal government could cut costs and become more efficient. The Grace Commission recommended privatizing numerous government activities and outsourcing certain support services to private firms.[82] This recommendation was put into practice alongside a dramatic increase in defense spending. In response, government subcontractors and the defense industry increasingly sought out space near the nation's capital, turning the remaining farms of northern Virginia into spaced-out office complexes. These new buildings brought the sheen of glass windows to the traditionally suburban and rural landscape. However, unlike the office buildings surrounded by concrete sidewalks and streets in most metropolitan centers, many of those in northern Virginia were surrounded on all sides by sixty feet or more of "mowed grass" and "a corridor of brushed raked dirt."[83] While the manicured lawns reflected green off the opaque windows of the office buildings and thus helped maintain the seemingly close-to-nature aesthetic of the area, such landscaping was not the result of a local zoning ordinance. Instead, it was a federal requirement placed on buildings that contain "sensitive compartmented information facilities" (SCIFs), a necessary component of the structures occupied by government subcontractors that handle large amounts of classified information. The open space around the buildings were supposed to make any potential perpetrators easy to see and the raked dirt paths made new footprints easy to detect.[84] While such lawns and paths were not particularly hospitable to ticks, they fragmented both the forests that dotted the land and halted the spreading concrete jungle of Washington, DC. As such, the SCIF landscapes of northern Virginia served as potential daily commuting corridors for people moving from their suburban acreages to work and for tick hosts scavenging for food and moving from one forest edge to another.

The building boom driven by government subcontractors and the defense industry in northern Virginia was not limited to commercial property. The surge in the availability of jobs also helped to fuel a new wave of migration to the area. In Fairfax County, the population increased from 596,901 in 1980 to 969,749 in 2000. To the northwest, Loudoun County's population increased from 57,427 in 1980 to 169,599 in 2000. To the south, Prince William County's population increased from 144,703 in 1980 to 280,813 in 2000.[85] Such population growth occurred despite the unraveling of New Deal housing and mortgage policies. While climbing mortgage lending rates at the time priced many people in the United States out of the housing market, the jobs created by subcontracted government service providers and defense contractors often paid very well. As a result, many of the new settlers of northern Virginia were still able to afford—and served as ideal borrowers—for higher-priced mortgages. Solidifying the existing sub- and exurban imaginary of the area, they frequently purchased or built sprawling residential acreages in seemingly serene and natural settings.[86]

In the late twentieth century, the ecosystem of northern Virginia thus was radically changed. Until the 1970s, the area was largely made up of agricultural and forested lands. While such land made good habitat for wildlife, the use of agricultural pesticides such as DDT (dichlorodiphenyl-trichloroethane) likely adversely affected the populations of birds and small mammals in the area. In addition, large mammals such as deer were sparse in northern Virginia due to centuries of overhunting. However, the late-twentieth-century human migrations to the area resulted in the conversion of agricultural land into residential estates and later coincided with the phaseout of DDT. In addition, the Virginia Department of Game and Inland Fisheries implemented a deer reintroduction program that put the mammal, but not its natural predators, into counties such as Fairfax, Loudoun, and Prince William where they were largely absent.[87] As the residential and office acreages of northern Virginia sprawled across the land, existing forests became more fragmented and new forest patches became more established. The new settlers of northern Virginia thus helped to create the ideal landscapes for the proliferation of both black-legged ticks' largest breeding grounds—on the coats of white-tailed deer—and Lyme diseases' most adept host—the white-footed mouse. Their large-lot residential neighborhoods surrounded by trees not only served as excellent food sources for deer and mice but also protected those mammals from being hunted by humans and natural predators alike. Deer in northern Virginia thus went from being nearly absent to abundant. Indeed, by 1980 Fairfax, Loudoun, and Prince William counties all had areas in which the deer population exceeded environmental capacity.[88]

These changes to the landscape of northern Virginia resulted not only in an increased presence of tick hosts but also in a greater abundance of blacklegged ticks and Lyme disease. Prior to the late 1980s, blacklegged ticks were not thought to exist in Virginia. Yet, by then, the incidence of Lyme disease diagnosed by doctors was on the rise. In 1989, the Virginia Department of Health began to require that medical professionals report cases of Lyme disease to the state.[89] In the first decade in which the disease was reported in Virginia, the annual average rate of confirmed cases hovered around 100. Those afflicted with Lyme disease, however, were disproportionately concentrated in northern Virginia. While 19% of the state's population lived in Fairfax, Prince William, and Loudoun counties,

28% of those who had Lyme disease lived in one of the three northern counties. From 2000 to 2009, the annual average number of confirmed cases of Lyme disease in the state increased significantly to 441.[90] The disparities in incidence rates in northern Virginia compared to elsewhere in the state also grew. While the three counties' combined population grew to make up 22% of the state's, 57% of those afflicted with Lyme disease lived in Fairfax, Prince William, or Loudoun County.[91]

## FINANCE, FOREST EDGE, AND THREE MAGICAL CREATURES

In Michael Farris's "northern invasion," blacklegged ticks were the "damn Yankees" that migrated to Virginia and caused undue harm to residents in the form of Lyme disease. But it was the invasion of the area by White European colonizers and their subsequent alterations to the local landscape that created the conditions for blacklegged ticks, their hosts, and Lyme disease to thrive. Like many of the people who reside in northern Virginia today, Michael Farris has a story of relocation to the area that is emblematic of just this trajectory. In the 1980s, Farris moved from Washington State to Virginia at a time when Lyme disease was rare and almost unheard of in the state. In 1987, he built a new, 3,918–square foot home on a three-acre lot in Loudoun County. His neighbors beside and behind him also built large homes on three-acre residential lots. Before being zoned for single-family residential development, the land on which Farris and his neighbors built was zoned for rural and agricultural uses. In addition, the land both across the road from the entrance to Farris's lot and touching a corner of the back of his property were open fields still being farmed. While a number of the newly constructed homes on lots near Farris's were completely enshrouded by trees, aerial imagery shows that Farris's lot was fairly barren and consisted mostly of open and well-manicured grass. The only trees present were in a thin strip at the front of the property along the road and a few small patches toward the rear.

However, the rustic rural and agricultural landscape on which Farris and his neighbors sought to build their early suburban homes no longer exists. In the mid- to late 1990s, the agricultural land in front of his house was slowly divided into residential lots, many of which were surrounded by conservation easements and lined with trees. At the same time, the farmland behind Farris's changed too. No new homes were built, but 22 acres of the land were planted with trees and turned into thick forest. In addition, in the early 2000s, a family from Maryland bought the land touching the back corner of Farris's lot and turned it into a 31-acre horse farm surrounded by small trees and shrubs.

The Farris family's bout with Lyme disease began in the mid-2000s. As the land near his house increasingly changed from agricultural to residential, more large lot homes dotted the landscape and created increasing amounts of forest edge. In Farris's own words, "Forest edge is ideal for growing the environment in which ticks, white-footed mice, and deer—which are kind of the three magical creatures—thrive. And suburban Virginia just creates infinite amounts of forest edge." Speaking about the wildlife around his own house, Farris stated that, almost every day, he had twenty or thirty deer in his yard between 6 and 7 a.m. that he would have liked to shoot if it would not endanger the neighbors.

Despite Farris's southern conservative credentials as a hunter and god-fearing proponent of small government, the form of his lot and the infill of the agricultural and timber fields near his own home were the result of direct government intervention. Changes to banking and mortgage policies in the United States in the 1980s that essentially handed control over housing to Wall Street facilitated the emergence of sprawling suburban estates like Farris's. Loudoun County zoning codes also favored large-lot residential development under the guise of sustaining a rural aesthetic in order to protect the property values of the White residents of what today is one of the wealthiest counties in the United States. And the migration of people with well-paying jobs to the area coincided with the massive shift toward the subcontracting of government services from the 1980s onward. Taking these facts into account, Farris and his neighbors are perhaps best seen as part of the invasion of northern Virginia that helped to turn the landscape into a more ideal environment for ticks, tick hosts, and tick-borne disease.

# 2 The Worst Animal in the World

In August 2020, Joshua Sokol published an article in *The Atlantic* titled "The Worst Animal in the World."[1] The article could very well have been about blacklegged ticks, but the animal under investigation, the *Aedes aegypti* mosquito, was portrayed to be much, much worse. Like the black-legged tick, the mosquito ingests, as its primary food source, fresh blood. In addition, it is known to cause an array of debilitating diseases. However, while blacklegged ticks could live in your backyard and feed on human blood by happenstance, *Ae. aegypti* could live in your house and feed on human blood with outright intentionality. Furthermore, while most of the diseases caused by blacklegged ticks are bacterial and can usually be treated using antibiotics, the diseases caused by *Ae. aegypti* are largely viral with no known treatment.

According to Sokol, not only was the *Ae. aegypti* "the worst animal in the world," but it had achieved "world domination." While its ancestral home has been traced back to the tropical and subtropical regions of Africa, over the past five hundred years it has established a presence on every continent except Antarctica. *Ae. aegypti*'s breeding and feeding habits well suited it to achieve domination on the world stage. The mosquito prefers to lay its eggs in artificial receptacles of water, and its offspring have been shown to emerge from containers as small as a bottle cap.[2] Its eggs can remain viable out of water for months.[3] And it prefers to feed on humans.[4] These characteristics have made *Ae. aegypti*'s presence in densely populated human settlements common and earned it the title the "urban mosquito."[5] Such traits have also made *Ae. aegypti* particularly adept at transoceanic travel. While the mosquito is unable to fly more than a few hundred feet, it can hitch a ride and survive on almost any vessel that carries water and people.[6]

In the past ten years, *Ae. aegypti* has made a somewhat regular appearance in the popular press. In 2015, the mosquito was identified as the primary vector for the outbreak of Zika in Brazil, a virus not previously known to exist in the country. But *Ae. aegypti* can actually carry more deadly diseases than Zika. Historically, it may be most famous for carrying a different arbovirus: yellow fever.[7] While vaccines made the disease less common, a serious bout of yellow fever can lead to organ failure and occasionally death. It is estimated that the virus still causes 30,000 deaths a year.[8] *Ae. aegypti* is also the primary vector of dengue. While most people recover from the virus, some cases lead to severe bleeding, organ failure, and plasma leakage.[9] Each year, dengue causes approximately 21,000 deaths.[10]

In Brazil, *Ae. aegypti* and the diseases it carries have been depicted most often as plaguing spaces of urban poverty. In examinations of Zika, scholars and journalists alike placed the initial epicenter of the disease in the northeastern coastal city of Recife.[11] The city's mosquito index demonstrated that mosquito larvae were three times more likely to be found in houses in the poorest neighborhoods than in the wealthiest ones. The homes in poorer neighborhoods had limited access to running water and were thus more likely to have tanks or barrels for water storage. In addition, trash pickup was less reliable and controlled stormwater runoff was often nonexistent.[12] Unsurprisingly, rates of Zika-caused microcephaly, a condition in which babies are born with smaller heads, in poorer neighborhoods were also significantly higher than in wealthier parts of the city.[13] Scholars examining dengue fever in Recife and in Brazil's other coastal cities have made similar observations about the presence of *Ae. aegypti* and likelihood for an outbreak of the disease in impoverished urban neighborhoods.[14]

Over the past three decades, however, incidence rates of the diseases commonly carried by *Ae. aegypti* in Brazil's more rural yet wealthier agricultural interior have frequently matched if not exceeded those in coastal urban areas. When the Zika virus first struck Brazil in 2016, most cases were reported in the more densely populated coastal states. By the end of 2017, though, Zika incidence rates in the agricultural state of Mato Grosso were several times higher than in Brazil as whole.[15] Similarly, when dengue first reemerged in Brazil in the 1980s and 1990s, most reported cases were isolated in the country's eastern cities. But, since 2005, the central west region, including Mato Grosso, has reported the highest incidence rates of dengue in the country almost every year.[16] But if what makes *Ae. aegypti* the worst animal is in part its urbanity, how did the pathogens the mosquito carries end up plaguing an area known more for its fields of soy than its impoverished slums?

Despite this seeming ecological incongruence, the presence and spread of both dengue and Zika to Mato Grosso do seem to support Sokol's notion about *Ae. aegypti's* potential for "world domination." However, like Michael Farris's "northern invasion" of Virginia by blacklegged ticks, Sokol's observations in some ways obfuscate how another quest for domination helped bring the mosquito and the diseases it carries to unexpected places such as Mato Grosso. The Cerrado of central Brazil was long eyed as an area of potential agricultural development but was deemed unsuitable because of the high levels of acidity and low levels of fertility of its soils.[17] While some believed that large agricultural enterprises could overcome such difficulties through economies of scale, the Brazilian state prioritized industrial development in its coastal cities. Through this process, the state created ideal environments for an "urban mosquito" and thus dengue and Zika. In the final decades of the twentieth century, however, the ways Brazil was incorporated into the global economy changed. Caught up in the global debt crisis of the 1980s, Brazil was forced by its international lenders to liberalize its economy. In the process, the country's agricultural sector became an attractive site of investment, and the landscapes of the Brazilian interior were radically transformed. As global finance provided the monetary backing to turn the tropical savannas of Mato Grosso and elsewhere into never ending rows of cropland, the quest of the country's agricultural sector to dominate the ecology of the Cerrado resulted in the creation of new environments in the Brazilian soyscapes that allowed *Ae. aegypti* to enhance its presence.

## THE EARLY HISTORY OF THE *AE. AEGYPTI* AND ARBOVIRUSES IN BRAZIL

The *Ae. aegypti* depicted by Sokol and others in the popular media is a relatively new incarnation of the pest. For most of human history, the mosquito did not necessarily like living in cities or feeding on the blood of humans. In its ancestral form, the mosquito is thought to have largely lived in forested areas where it preferred to feed on nonhuman animals and breed in tree holes.[18] Mosquito scientists believe that in adjusting to periods of prolonged drought, a subspecies of *Ae. aegypti* evolved and became "domesticated." Phylogeographic studies demonstrate that this subspecies of the mosquito likely originated in western Africa 400–550 years ago. Shortly after its domestication, *Ae. aegypti* is believed to have made its way to the Americas. Once isolated across the Atlantic, *Ae. aegypti* evolved independently of its forest-dwelling relative. Today, two known subspecies of *Ae. aegypti* thus exist: the urban "domesticated" *Ae. aegypti aegypti* and the sylvatic, "wild" *Ae. aegypti formosus*.[19]

*Ae. aegypti* prefers to live in places with warmer temperatures and higher levels of precipitation. The mosquito is most commonly found where temperatures do not drop below 15° Celsius or rise above 30° Celsius.[20] While it can live at higher temperatures, the mosquito is debilitated by, and unlikely to survive exposure to, lower temperatures.[21] *Ae. aegypti* can live in dryer environments if it has access to a water and food sources. However, places with higher levels of precipitation often provide the mosquito with more potential sites to lay its eggs. Both male and female *Ae. aegypti* require a meal after reaching adulthood and before breeding. Male *Ae. aegypti* obtain their food largely from sugary plant sources, such as flower nectar.[22] They do not bite and thus pose little threat to humans or other animals. Female *Ae. aegypti* occasionally feed on nectar as well. However, they require a bloodmeal to produce eggs. Although female *Ae. aegypti* feed on other animals, they are attracted to the smell of humans and feed on them at a much higher frequency.[23] *Ae.*

*aegypti*'s feeding behavior makes it particularly well suited to spreading infectious diseases. While most mosquitoes secure a bloodmeal with one bite, *Ae. aegypti* is a sip feeder and thus takes little sips of blood from multiple hosts in order satiate its nutritional needs.[24]

The spread of *Ae. aegypti* and the pathogens it carries is more than a story of evolution and adaptation. The mosquito's worldwide presence is intimately linked to the expansion of global capitalist trade routes. Sometime in the sixteenth century, the *Ae. aegypti* is thought to have boarded Portuguese slave ships in Angola and made its way to the coastal port cities of Brazil and northern Argentina.[25] Innumerable enslaved African people died in the passage, victims of the inhumane treatment of their captors.[26] However, with a readily available supply of human blood to feed on and containers of freshwater to breed in, *Ae. aegypti* were able to live and reproduce during the transatlantic voyage.[27] Enslaved Africans were often blamed for carrying yellow fever, but given the endemic status of the virus in West Africa, they likely had acquired immunity from prior infection before making the voyage.[28] The virus was thus likely transported overseas by the ships' European crews and *Ae. aegypti* itself.[29] Dengue is also thought to have first arrived in Brazil during the colonial period. However, its origins and immediate impact are not as well traced.[30]

As the altered landscapes of North America affected the presence of blacklegged ticks, the altered landscapes of Brazil similarly impacted the presence of mosquitoes. Prior to the arrival of Portuguese colonizers, an extensive evergreen, moist forest lined most of the Brazilian coast. The drier savannas and grasslands of the Cerrado separated the coastal Atlantic forest from the Amazon, but some riparian forest corridors still connected the two tropical areas.[31] As elsewhere in the world, humans had long shaped this landscape. It is estimated that by the time the Portuguese arrived, 60% to 80% of Brazil's forests had been modified in some way.[32] Brazilian Indigenous groups are known to have engaged in *coivara*, a slash-and-burn form of agriculture used to increase soil fertility. After burning, the land could be used for about three years. While new forest was then cut and burned, the area left fallow is thought to have been tended in order to encourage the growth of some trees and plants over others.[33] These practices likely limited the presence and prevalence of some diseases in precolonial Brazil. Overall, the amount of forest being used was relatively small in scale, and the forest was allowed to regenerate. In turn, *coivara* practices likely had less of an impact on biodiversity, in particular that of mosquito predators such as birds and lizards.[34] In addition, Indigenous populations were not thought to have kept domesticated animals, making animal-to-human disease spillovers uncommon.[35] The overall lack of deadly disease such as malaria found in other parts of the world led Portuguese explorers to see Brazil as endowed with "good air," and they frequently extolled the health of Indigenous populations and the seeming lack of "pestilence or illness."[36]

However, Portuguese colonizers not only brought *Ae. aegypti* and the diseases it carried to Brazil but also created landscapes on which it could thrive. With the establishment and expansion of sugar plantations in the seventeenth century, the Portuguese felled massive swaths of the Atlantic forest. To increase soil fertility and open land to cultivation, plantation owners frequently adopted the slash-and-burn methods used by Indigenous populations. However, engaging in large-scale nutrient-intensive monocrop production, plantation owners continually burned more forest to bring new land under cultivation.[37] Furthermore, plantation owners did not seek to fallow the land and return it to production. Instead, they frequently used abandoned fields as pasture. With cattle and other animals grazing on sprouted saplings and other native plants, the forests and wildlife that lived there were unable to regenerate and thus were seceded by grassland species.[38] Processing sugarcane into sugar also further drove deforestation. It is estimated that each sugar harvest used between 4 and 22 hectares of timber to make charcoal to fire the furnaces and make the crates to transport the finished product.[39]

Plantations provided *Ae. aegypti* with excellent places to feed and breed. The plantation system depended on large amounts of Indigenous and African slave labor to plant, harvest, and process the sugarcane. With a concentrated supply of human blood, female *Ae. aegypti* could easily acquire the necessary amount of food to satisfy their nutritional needs. Deforestation is also thought to have reduced mosquito predators such as insectivorous birds and reptiles.[40] *Ae. aegypti* could thus more easily mature to an egg-laying age. In addition, sugarcane processing entailed drying a sweet liquid into a solid in hundreds if not thousands of clay pots, providing male *Ae. aegypti* with a viable food source. Only used a few months of the year and prone to break, the pots and their fragments also frequently filled with rainwater in the tropical environment. As a result, the *Ae. aegypti* had easy access to an array of potential egg-laying sites.[41]

Sugar production led to the creation of an ideal environment not only for *Ae. aegypti* but also for the pathogens it could carry. Jesuit missionaries reported on, and likely brought, diseases such as smallpox and influenza to Indigenous populations as early as the mid-sixteenth century.[42] The sugar trade completed the Atlantic triangle, linking people, pests, and pathogens on three continents. Coastal port cities served as the initial places where diseases such as yellow fever and dengue could spread. Carried by the crew members of slave ships and the mosquito itself, the viruses could easily disembark into ports full of nonimmune Indigenous populations and Portuguese colonizers.[43] As transfer points for the buying and selling of processed sugar and African slaves, the port cities were sites where the viruses could infect a plantation owner or overseer, who would then take them beyond the city. On the plantations, they could then spread among nonimmune Indigenous slaves and Portuguese colonizers. Indeed, by the mid-seventeenth century, the diseases brought by the Portuguese to Brazil had decimated Indigenous populations, leaving tens, if not hundreds, of thousands of people dead.[44]

In the eighteenth and nineteenth centuries, Brazil's export and agricultural activities shifted to include coffee, cattle, and other food crops, but the movement and mingling of people, pests, and pathogens continued and expanded largely uninterrupted. Having turned what was once a biodiverse forest, running 2,300 kilometers along the coast and extending 100 kilometers inland, into fields, pastures, or land degraded beyond use, Portuguese colonizers created ideal conditions for the spread of *Ae. aegypti* and diseases such as yellow fever and dengue.[45] While infectious disease outbreaks occurred with less frequency than in the Caribbean, Central America, and North American, by 1850 outbreaks had dramatically increased. In 1887, the *Rio News* even referred to the country as "nothing less than a huge pest-house where . . . contagious diseases are constantly in existence."[46]

Over the following decades, the Brazilian state sought to end the perception that the country was a place of pestilence. To do so, the state gradually sought to create landscapes in which the diseases plaguing the country were thought to be less likely to thrive. Having control only over places under federal jurisdiction, the state initially focused on improving public health in its ports and the capital of Rio de Janeiro. Drawing on Cuban scientists' recent observation that the presence of mosquitoes was linked to outbreaks of yellow fever, Brazilian public health officials sought to eliminate the places where *Ae. aegypti* lived and bred. This meant modernizing Rio de Janeiro's infrastructure by widening streets to increase ventilation and installing proper sewage, water, and drainage systems in the city's central areas. The process entailed the demolition of 590 buildings, including many of the city's overcrowded tenements. It also led to creation of sanitary brigades, known as *mata mosquitos* or "mosquito killers," that went door to door in the city to find and eliminate *Ae. aegypti* through fumigation methods. In 1909, public health officials declared their efforts a success and claimed to have eradicated yellow fever from Rio de Janeiro.[47]

Although Brazil's early efforts to stymie disease were framed as part of a nation-building and modernization process, they were largely shaped by the interests of the Brazilian elite. In Rio de Janeiro, the demolition of tenements resulted in the displacement of working-class families from the city center and into the favelas lining the city's hillside. Efforts to eliminate yellow fever in Rio de Janeiro thus occurred alongside efforts to eliminate the presence of most of the working-class population from the traditional inner city.[48] At the same time, the Brazilian elite in part accepted federal-led interventions to control yellow fever because they believed the country's image as a place of sickness was a deterrent to attracting European immigrants. For the elite, such immigrants served not only to bolster the available supply of labor but also to whiten the general population. Many local elites held the racist belief that Brazil's majority Black and Indigenous population and a lack of White European immigration was stymying the country's economic growth and preventing it from becoming a "civilized nation." With European immigrants making up the majority of people dying in Brazil from diseases such as yellow fever, public health was linked to elites' desires to rescue "Brazil's European legacy" from the "contamination of inferior cultures."[49]

The acceptance and success of measures to control yellow fever and other diseases gradually spread outside Rio de Janeiro, but federal intervention was largely isolated to coastal cities. Attempts by public health officials to expand their efforts to agricultural areas were constrained by elites' concerns of being subjected to federal oversight. As a result, large landholders frequently refused to allow the federal government to assist in efforts to combat yellow fever on their property.[50] Spatially vast and difficult to access, the Brazilian interior had also long been neglected by federal policy makers. In the 1910s, doctors and researchers began to note the grave differences in the health and livelihoods of the country's interior rural and coastal urban inhabitants. A Brazilian physician named Miguel Pereira referred to the Brazilian interior as a "vast hospital."[51] The report underpinning Pereira's observation detailed that ill-health was not an inherent attribute of an individual or group of individuals but linked to poverty and a lack of access to necessary resources. This revelation resulted in the creation of some rural health outposts. However, the Rockefeller Foundation's entrance into the country's public health arena in 1917 led in part to a doubling down on the federal state's focus on yellow fever in urban areas. Despite observations by Brazilian doctors and scientists that yellow fever afflicted communities in rural areas, the Rockefeller Foundation's focus solely on the elimination of the disease's vector—*Ae. aegypti*—led it, in the words of foundation director Joseph White, to "pay no sort of attention to the small towns."[52] Such efforts did lead to the eradication of yellow fever in Brazil's urban coastal centers, but the eradication of yellow fever in the nation as a whole remained highly elusive.[53] Whereas the presence of *Ae. aegypti* and the diseases it carried in Brazil from the seventeenth through the early twentieth century were heavily shaped by the material desires and legacies of the country's colonizers, from the mid-twentieth century onward, the forces that most heavily shaped where in Brazil the mosquito would inflict its toll were the internal and external flows of finance within and to the country.

## THE SPILLOVER: FINANCE, LANDSCAPE CHANGE, AND THE REMAKING OF MOSQUITO ENVIRONMENTS

Efforts to improve public health in early-twentieth-century Brazil were based on the local elite's racist beliefs that linked economic growth to Whiteness, but Pereira and his colleagues laid the foundations for the

Brazilian state to eventually act on the connections between moderniza-tion, social inequality, and public health. In accord with Brazil's develop-mental state-building policies of the mid-twentieth century, the state used the auspices of public health and the centralized distribution of finance to shape local landscapes and justify its expenditures. In the process, a number of infectious diseases—including yellow fever—were temporarily eradicated in some parts of the country. However, as Brazil fell victim to the global debt crisis in the late 1970s and early 1980s and was forced to adopt structural adjustment policies, the state ceded its power to direct the flows of global finance coming into the country. As it did, public health no longer served as a pillar of modernization and development, and the country's landscapes once again became places where pests and patho-gens could thrive.

In the 1930s and 1940s, the Brazilian state worked to improve its institutional capacity to expand public health efforts throughout the entire country. Although such efforts had begun decades earlier, the Ministry of Education and Public Health was created in 1930 to address health in the nation as a whole. The ministry was at first underfunded but worked with the Rockefeller Foundation to extend its reach. Among the partnership's notable achievements was the elimination of *Anopheles gambiae*, the pri-mary vector for malaria, in northeastern Brazil by 1941.[54] In addition, the partnership's efforts were thought to have eradicated *Ae. aegypti* in much of Brazil during the 1940s.[55] Within the Ministry of Education and Public Health, agencies were also created to combat specific diseases, including the National Yellow Fever Service (Serviço Nacional de Febre Amarela) and the National Malaria Service (Serviço Nacional de Malária).[56] In 1942, the Special Service for Public Health (Serviço Especial de Saúde Pública) was created as a joint project with the Institute of Inter American Affairs, an agency within the US State Department. While the entirety of Brazil fell under the jurisdiction of the National Yellow Fever Service and the National Malaria Service, the Special Service for Public Health was in charge of controlling disease in sites of strategic natural resource extrac-tion, in particular rubber and iron ore, deemed essential to the US war effort.[57] These services targeted different diseases and vectors, but their tactics often had overlapping effects. Most notably, their widespread use of larvicides and pesticides, in particular dichlorodiphenyltrichloroethane

(DDT), impacted different insect species indiscriminately, killing yellow fever and malaria mosquito vectors alike. The use of chemical warfare as a public health measure led to a decline in mosquitoes and mosquito-born disease in some parts of Brazil, but shifts in the ways the country was incorporated into the global economy in the postwar era perhaps had an even greater impact on the health of the country's people.

In 1953, the Joint Brazil–United States Economic Development Commission (JBUSEDC) concluded that sustainable economic growth in the country could be achieved only by overcoming so-called "infrastructural bottlenecks." In response, the government put forward a series of plans to direct public investment into energy provision, transportation, and communication networks and to subsidize "germinative areas" of the private sector.[58] In 1952 the JBUSEDC had recommended that Brazil create the National Bank for Economic Development (Banco Nacional de Desenvolvimento Econômico, or BNDE) to fund such plans. The BNDE was to be funded by the JBUSEDC and a 15% tax levied by the state on the net earnings of large corporations. However, shortly after the release of the JBUSEDC's 1953 report, the United States sharply decreased its lending activities to Latin American countries and unilaterally discontinued the joint commission's work. As a result, a key source of funding for Brazil's development plans disappeared, and its modernization and industrialization programs stalled.[59]

Brazil's push for development was revitalized in 1956 with the introduction of what became known as the Targets Plan (Plano de Metas). With the stated goal of achieving fifty years of economic growth and development in five years, the Targets Plan built on the JBUSEDC plan to rapidly develop the country's energy, transport, basic industries, food, and education sectors.[60] To overcome the difficulties of funding the program, the government leaned heavily on the Bank of Brazil (Banco do Brasil), the Superintendency of Money and Credit (Superintendência da Moneda e do Crédito), and the National Bank for Economic Development to provide loans and devise policies that would foster the development of public and private enterprises and the buy-in of domestic and foreign investors. The Bank of Brazil and the Superintendency of Money and Credit used their control over imports and exports to levy tariffs and control foreign exchange in ways that met the Targets Plan's goals. Perhaps most notably, in 1955 the Superintendency of Money and Credit issued Instruction 113, a measure that permitted foreign private investors to import certain types of capital equipment without having to go through exchange controls. This allowed companies to buy and import equipment with their own supplies of foreign currency, in particular dollars, thus avoiding having to purchase Brazilian cruzeiros at the market rate and then repurchase dollars at a higher price.[61] The Bank of Brazil's foreign trade department was responsible for approving companies to use Instruction 113 and also held a monopoly position over the buying and selling of foreign currency.[62] The National Bank for Economic Development was largely responsible for funding or incentivizing investment in the strategic sectors identified in the Targets Plan. While the United States never fulfilled its funding pledge to the National Bank for Economic Development, the Brazilian state levied an array of surtaxes whose revenues went directly to the bank. Once the surtaxes were approved by the Brazilian Congress, they no longer required annual approval and thus proved to be somewhat insulated from politics and in turn a more stable funding stream.[63] In total, the primary agencies overseeing the funding of the Targets Plan were able to mobilize close to US$7 billion, approximately 14% of Brazil's GDP at the time.

While the Targets Plan goal was to modernize Brazil through rapid economic growth, the mechanism by which the government directed funds to achieve this goal would alter the landscapes and the distribution of vector-borne disease in the country. Overall, the vast majority of funds and benefits from the Target Plan went toward industrialization of the Brazilian economy. Foreign manufacturers that had dollars to purchase equipment that could then be imported into the country gleaned the greatest benefits from Instruction 113.[64] In addition, from 1957 to 1963, 48.6% of loans from the National Bank for Economic Development went toward basic industry and an additional 40.8% went toward the energy sector to fuel it.[65] Since Brazil's industries were already established in or near its southeastern coastal cities, in particular São Paulo, such places received the greatest benefits from the Targets Plan. From 1950 to 1960, São Paulo's share of Brazil's industrial output rose from 46.8% to 55%.[66] By the early 1960s, this industrialization had turned Brazil into a majority-urban society. Offering the greatest economic opportunity, the cities in the southeast region received the largest number of internal migrants.[67]

With Brazil's rapid urban migration boom came the proliferation of the unplanned settlements known as favelas. As noted earlier, favelas emerged around Rio de Janeiro in the early 1900s, after the federal government tore down the city's tenement housing near the ports in order to modernize the city and make it more sanitary. This did contribute to the eradication of yellow fever in Rio de Janeiro's center, but it also displaced the working-class population to the hills lining the city. With more people moving to the city in the mid-twentieth century, the populations in the hillside favelas continued to mushroom.[68] Mass migration to São Paulo led to the formation of favelas there also, though construction of public low-income housing tempered their number and size. Other parts of the country faced similar housing situations. Favelas are thought to have housed the majority of the population in some poorer regions of the North and Northeast.[69] By 1964, estimates placed the population living in the favelas of Rio de Janeiro between 430,000 to 1,070,000 and of São Paulo between 50,000 to 100,000.[70] In the northeastern cities of Fortaleza and Recife, people living in the favelas are thought to have made up 25–50% of the population.[71]

Occupying marginalized spaces in Brazil's urban landscape, favelas proved to be a favorable environment for the *Ae. aegypti* mosquito and the diseases it carried. In and around Rio de Janeiro, the most desirable locations to build homes on or near the coast or in the valleys were often already owned and occupied. Favelas thus sprouted up on and scaled the surrounding hillsides.[72] Elsewhere in the country, favelas were constructed near or on the edges of rivers or roads on government-owned land.[73] Since they were subject to landslides and flooding and lacked access to running water, sewage, or sanitation services, standing water in the favelas was commonplace. Although Brazil's mosquito eradication programs included some of these places, it missed many. Favelas were thus ripe environments for *Ae. aegypti*.

In the 1960s, the Brazilian state began to take more aggressive measures to eliminate the favelas. Like housing reformers who sought to eliminate tenements in US cities in late nineteenth and early twentieth centuries, Brazil's new military dictatorship and local elites saw the favelas as breeding grounds for disease and discontents alike. To eradicate such perceived problems, the state enacted Law 4.380, which implemented a new housing and banking system, that in some ways resembled the one created during the New Deal in the United States. Most notably, the law resulted in the creation of the Housing Finance System (Sistema Financeiro de Habitação), the National Housing Bank (Banco Nacional da Habitação, or BNH), the Guaranteed Employment Fund (Fundo de Garantia do Tempo e Serviço), and the Brazilian Savings and Loan System. The stated goal of the Housing Finance System was "to promote the construction and acquisition of homes, especially by the lower-income classes."[74] To reach this goal, the National Housing Bank would lend money to regional, state, and municipal agencies charged with addressing local housing problems. The BNH was funded through the Guaranteed Employment Fund and the Brazilian Savings and Loan System. The Guaranteed Employment Fund required employers to contribute 8% of their payroll to the public unemployment and retirement system, and the BNH was able use some of these funds as loans. The Brazilian Savings and Loan System operated like S&L banks in the United States, offering higher interest rates to persuade everyday people to deposit their savings. The Brazilian Savings and Loan System and other National Housing Bank loans were indexed to the rate of inflation but could not exceed an individual's rate of wage increase.[75] The National Housing Bank could also raise funds through the sale of inflation-indexed real estate bonds.[76]

While the Housing Finance System was created to improve the living conditions of low-income Brazilians, wealthier city dwellers may have benefited more from its various programs. The National Housing Bank indirectly created employment opportunities for lower-income Brazilians by funding residential and infrastructural construction projects. Some local programs also helped favela residents to secure legal title to the land they occupied and extended infrastructure such as water, sewage, and electricity to their homes. In addition, BNH loans were used to help developers construct, and individuals purchase, millions of houses and apartments.[77] However, the construction of this housing was often used to justify the demolition of favelas near wealthier urban enclaves or on profitable land wanted for other uses. The housing was also frequently built far from city centers, with building materials of questionable quality and in inadequate numbers of structures and units to meet demand. As a result, not all displaced favela residents were able to acquire a new home, and those who did often had a much longer daily commute.[78]

In the 1970s, it became apparent that the National Housing Bank's lending programs were largely inadequate for addressing the housing situation of lower-income Brazilians. Declining real wages made it increasingly difficult for them to obtain and pay off loans.[79] Numerous favela residents did take out housing loans to take advantage of the lower interest rates for less expensive homes offered by the National Housing Bank. However, whereas the homes they constructed in the favelas often had no monthly payment, the homes constructed for them with BNH funds did. In addition, while wages were usually adjusted annually to account for inflation, the National Housing Bank inflation-indexed mortgages were adjusted every three months.[80] While safeguards were in place to protect borrowers from such discrepancies, they often did not align fast enough with people's financial realities. Facing economic hardship, many borrowers thus defaulted on their loans. In some major urban areas, default rates surpassed 70%.[81] In 1975, the BNH recognized that one-third of urban families did not have the financial means to participate in its existing housing programs.[82] Thus, despite aggressive campaigns to eradicate the favelas, they continued to grow.[83]

The National Housing Bank's investment portfolio and income sources also began to change. Most notably, the BNH was transformed from merely a mortgage lender to an urban development bank. From 1969 to 1978, home loans declined as a proportion of annual lending from 93.2% to 66.9%. At the same time, from 1969 to 1974, the National Housing Bank's investments in urban development jumped from 4.1% to 28.1% of annual lending.[84] This diversified the National Housing Bank's portfolio to include seemingly more reliable borrowers such as commercial entities and state and municipal governments. However, as economic stagnation began to more severely impact Brazil in the latter half of the 1970s, the income from these borrowers was not enough to make up for the losses to the National Housing Bank's more traditional funding streams. In addition to lost revenue from defaults, funds acquired through the Guaranteed Employment Fund's payroll payments decreased with dips in Brazil's economic activity. At the same time, as individuals fell into economic hardship, they withdrew more funds, and deposits declined in the Brazilian Savings and Loan System.[85] Since its founding, the National Housing Bank's lending activities had been financed almost completely by its own revenue streams and its originally allocated resources from the Guaranteed Employment Fund and the Brazilian Savings and Loan System. As such funding shrank, the National Housing Bank began seeking out funding from external sources.

With the world awash in petrodollars, the National Housing Bank—and the Brazilian government more broadly—was able to continue to fund its projects and national development by taking on more debt. Seeking out investment opportunities, international banks actively sought to lend money to those states and other government entities that they perceived to be more secure borrowers. As a result, Brazil dramatically increased its acquisition of funds from international lenders in the 1970s. At the beginning of the decade, Brazil's total foreign debt sat at approximately US$5.5 billion. By 1980, that figure had increased to US$60.8 billion and would continue to rise by nearly US$10 billion annually until 1985.[86] As for many developing countries in the world, the loans extended to Brazilian borrowers had flexible interest rates that were initially very low. In addition, most loans were made in US dollars. As the United States sought to lower the supply of US dollars in the global marketplace in the late 1970s, international lending rates increased and exports from countries such as Brazil decreased. From 1978 to 1981, Brazil's annual net interest payments on its foreign debt increased from US$2.7 billion to US$9.2 billion. In addition, the country's terms of trade (the price for its exports relative to that for imports) declined by 30%.[87] In December 1982, Brazil turned to the International Monetary Fund (IMF) for assistance. With the IMF's backing, international banks agreed to renegotiate Brazil's loans on condition that the country liberalize its economy.[88]

The National Housing Bank was somewhat better able to continue its lending activities, largely with internal funds, longer than other federal government entities. To receive IMF support, however, the Brazilian state was required to decrease social spending and phase out federal support for most public sector housing programs.[89] Initially, the BNH was still able to fund some of its infrastructural programs with the help of foreign investors. From 1981 to 1987, the funds for BNH projects coming from external investors rose from 1.3% to 13.5%.[90] In the process, the National Housing Bank became one of the top recipients of World Bank loans worldwide.[91] Such funds were not enough, however, to compensate for the bank's loss of revenue because of smaller contributions to the Guaranteed Employment Fund and more numerous loan defaults. As the National Housing Bank's lack of funds began to prevent it from performing its basic duties, it became a popular target among politicians. On November 21, 1986, the National Housing Bank was abolished.[92]

Brazil's adoption of austerity measures and the end of the National Housing Bank were accompanied by a dramatic rise in poverty and a restructuring of urban space. The poverty rate increased from 24.3% in 1980 to 39.3% in 1988.[93] At the same time, state support of housing finance and changes in its regulation resulted in the increased densification of Brazil's urban core. As in the United States, shifts in housing finance in the 1980s made it more difficult for less affluent segments of Brazilian society to acquire loans, and real estate investment and new residential construction became more concentrated in the most affluent areas of its coastal cities.[94] Favelas expanded as well. From 1980 to 1991, an estimated 105 new favelas emerged in Rio de Janeiro, raising the total to 564. At the same time, existing favelas expanded in size and became more densely populated.[95] The favela population also dramatically increased in São Paulo from 439,721 people in 1980 to 1,901,893 in 1993.[96] In northeastern cities such as Recife, the favela population grew to nearly 60% of the population, with 1,472,202 living in squatter settlements by the early 1990s.[97]

The expansion and densification of Brazil's urban populations also created conditions ripe for the return of *Ae. aegypti*. Some of the National Housing Bank infrastructural programs in the late 1970s and early 1980s brought water, sewage, and sanitation services to the favelas.[98] However, the expansion of such services was largely unable to keep up with the influx of new people. In addition, such services did not extend to newer favelas. Seeking a way to store water for everyday use, many residents resorted to having water storage containers in their homes. A survey of Brazilian homes in the 1980s found that 33.8% lacked a clean water supply and 39.8% had no garbage collection.[99]

With smaller water receptacles and puddles of stagnant water readily available in Brazil's urban areas, *Ae. aegypti* was able to make its return. The mosquito first reentered Brazil in 1967 in the northern city of Belém, but aggressive eradication efforts enabled the Pan American Health Organization to again declare the country *Ae. aegypti* free in 1973.[100]

Later in the 1970s, however, *Ae. aegypti* was found in Rio de Janeiro and Salvador, Bahia. The reappearance of the mosquito in Brazil's larger eastern coastal cities during a time of political and economic turmoil made the revival of nationwide eradication efforts difficult. In the 1980s, the *Ae. aegypti* gradually reestablished itself in Brazil as a whole. In turn, localized and seasonal outbreaks of dengue began to occur again in Brazil. In 1981, a major outbreak occurred in the Amazonian state of Roraima. In 1986 and again in 1990, major outbreaks occurred in the city of Rio de Janeiro, each resulting in more than 100,000 reported cases.[101] By the mid-1990s, dengue had become endemic in the country, occurring both during and outside the rainy season.[102] Increases in the reported cases of dengue continued into the twenty-first century, with incidence rates increasing from 79.6 occurrences per 100,000 inhabitants in 2000 to 826 occurrences per 100,000 in 2015.[103]

The increasing presence of *Ae. aegypti* in Brazil's urban areas not only contributed to the increasing rates of dengue in the country but also facilitated the spread and establishment of Zika. Phylogenetic studies date the virus's arrival in Brazil to 2013.[104] The first known outbreak hit northeastern Brazil in late 2014 in or near Recife and spread southeast to Brazil's more populated cities of Rio de Janeiro, Salvador, Fortaleza, and São Paulo.[105] In 2016, nearly 50% of reported Zika cases were concentrated in Rio de Janeiro and Bahia.[106] As with dengue, caseloads were higher in urban impoverished areas where lack of access to water, sewage, and sanitation services often increased the presence of standing water and, in turn, *Ae. aegypti*.[107] Neither dengue nor Zika remained a disease of the urban poor alone, however. As Brazil's coastal urban areas were transformed by shifts in global finance, so was the country's more rural interior. In turn, the diseases once thought to plague only the urban poor spread to seemingly more unusual places.

## THE WORST ANIMAL IN THE BRAZILIAN AGRICULTURAL INTERIOR

In the colonial gaze, Mato Grosso has long been seen as an empty frontier to be conquered and transformed. European colonizers initially traversed the region in search of gold and Indigenous populations to enslave. Centuries later, Theodore Roosevelt revitalized the vision of Mato Grosso as a resource frontier after taking a trip down the Duvida River in the state's northeastern section.[108] However, Indigenous people long lived on and transformed the land. As in the United States and in the coastal areas of Brazil, White colonizers decimated Indigenous societies by introducing new diseases, enslaving some of the population, and rupturing the land's ecological balance.[109] The Portuguese crown parceled out land in Mato Grosso as political favors and with the hope that Europeans would settle the territory, but much of the land in the state remained unoccupied by the colonizers.[110] Fearing that some of its land might be taken by its southern neighbor in the mid-nineteenth century, the Brazilian state promoted the settlement of southern Mato Grosso (what is now Mato Grosso do Sul) after the Paraguay War. In the mid-twentieth century, the state more actively encouraged the settlement of the region through subsidized colonization programs and the expansion of transportation infrastructure.[111] But it was not until the late 1980s that the radical alteration of the central Mato Grosso landscape began and the region in turn became a more amenable home to the "worst animal in the world."

The Cerrado, the vast region spanning central Brazil and a significant portion of Mato Grosso, is one of the largest and most biologically rich tropical and subtropical savanna regions in the world. As a mosaic of grass and woodlands, it is home to more than 10,000 endemic plant species, close to 1,000 species of birds, and hundreds of different mammals, reptiles, and fish.[112] The region is sometimes referred to as the "birthplace of waters," and rivers originating there provide water to a significant portion of Brazil.[113] However, the Portuguese word *cerrado* translates to "closed." Despite the allure of Mato Grosso as a potential land of vast resource wealth, the Cerrado's infertile soils long made it useless for sustained agricultural production. The soils across much of the area are highly acidic and contain extremely low levels of phosphorus. In addition, their water-holding capacity is low, capable of maintaining a crop in full growth only for six to ten days without rain or irrigation.[114] These conditions, combined with a lack of infrastructure for transport, kept the Cerrado of Mato Grosso largely peripheral to Brazil's development until the mid-twentieth century.

In the 1950s, the Brazilian state began to appropriate and redistribute large swaths of Indigenous, peasant, and publicly held land. As it did, Mato Grosso became the site of some of the first private "colonization" companies. The state granted control over more than 4 million hectares of land to such companies, each of which then split them into twenty plots of 200,000 hectares each with the understanding that every plot would be cultivated or used for grazing.[115] Efforts to better incorporate Brazil's interior were further elaborated after the military coup in 1964 with the creation of the National Integration Program (Programa de Integraça Nacional, or PIN) and the National Institute for Colonization and Agrarian Reform (Instituto Nacional de Colonização e Reforma Agrária, or INCRA). The National Integration Program sought to integrate the Brazilian frontier with the coastal regions, most notably through the construction of the Trans-Amazonian and Cuiabá-Santarém highways.[116] The goal of the National Institute for Colonization and Agrarian Reform was to settle 100,000 landless families in the Brazilian interior within five years.

In Brazil's broader plans to "modernize" its economy, agricultural development remained largely peripheral, but some of the funds were directed toward programs such as the PIN, the INCRA, and other, more local agricultural and regional development agencies.[117] These agencies often encouraged investment of these moneys in the interior through substantial subsidies and generous tax deductions. From 1965 to 1975, the total amount of loanable funds available from the National System of Rural Credit increased more than fivefold. Starting in 1970, a sharp rise in inflation coupled with fixed nominal rates on agricultural loans resulted in years of negative interest rates for borrowers purportedly engaging in agricultural activities.[118] Benefiting from laws that also allowed companies to deduct half the amount they invested in infrastructure, industrial, or farming projects from their taxable income, the interior of Brazil appeared to be flush with capital during the 1970s.[119]

Brazil's efforts in the postwar era to turn its interior into an engine of economic growth proved to have limited success. In Mato Grosso and other agricultural areas, infrastructural development and aggressive land-titling efforts did occur. The portions of BR 163, the Cuiabá–Santarém Highway, that crossed the Cerrado in Mato Grosso were completed in 1975. In addition, 75% of all land in Mato Grosso was formally titled by

1980.[120] The construction of roads opened up new areas to potential development, but the occupation of the land was slow. Despite the National Institute for Colonization and Agrarian Reform's push to place 100,000 new families in the Brazilian interior, the agency succeeded in settling only 6,000 families in the area during its proposed five-year time frame.[121] Much of the redistributed public land was instead obtained by large landholders for speculation, as a means to secure a tax break or to access cheap credit that was frequently funneled back to the city to fund industrial development.[122] As a result, the land in the Brazilian interior became titled and owned but was often left idle.

The international debt crisis and the liberalization of the Brazilian economy in the 1980s did not impact just its coastal urban areas. The IMF's prescribed programs laid the foundation for the radical transformation of Brazil's agricultural sector and in turn the landscapes of its interior. Lacking funds, the government began to reduce the volume of agricultural credit and phase out agricultural subsidies in the early 1980s. By the end of the decade, the government had removed export restrictions and eliminated a number of export and import taxes and tariffs on agricultural products.[123] A constitutional amendment ratified in the mid-1990s eliminated any legal differentiation between Brazilian and non-Brazilian companies, allowing foreign firms to invest in sectors previously monopolized by domestic entities.[124] Although restrictions on rural land ownership remained in place, the changes allowed transnational firms to make inroads into Brazil's agricultural input, storage, and transport sectors. A series of laws was also enacted that allowed for the widespread financialization of the agricultural sector. In 1994, the Brazilian government created its first financial product that allowed farmers to acquire liquidity at the time of seeding through advanced sales of their projected harvests. In 2004, a number of additional financial instruments were rolled out that enabled producers to secure credit from a broader array of actors in the agribusiness supply chain. All of these financial instruments were securitized and made available in the global marketplace. By 2010, the market for agriculturally based financial products from Brazil topped US$140 billion.[125]

The post–debt crisis transformation of Brazil's agricultural sector solidified the dominance of large export-oriented farms. Given the greater use of land as collateral, banks were often more willing to extend credit to large farms. After the source of agricultural credit diversified to include other private actors in the agribusiness sector, large farms were still seen as the most creditworthy. Large farms could thus more easily take out loans to finance the purchase of the technologies—from tractors to seeds and fertilizers—necessary for large-scale export agriculture.[126] The broader financialization of the agricultural sector thus worked in the favor of large farms. On the one hand, large farms often had more capital available to hedge their risks in the futures market.[127] On the other hand, some large farms existed as agribusiness corporations whose shares could be bought and sold. The value of such farms was thus not only be linked to the land they owned and the crops they produced, but also to the demand for their shares on the stock exchange.[128]

The rise of large export-oriented farms in Brazil not only changed the country's agricultural sector but also ruptured the interior's ecology. As the planting of export crops such as soy, corn, and cotton dramatically increased, agricultural production expanded toward the center of Brazil. In the process, large amounts of Brazil's tropical savannah (the Cerrado) and rainforest (the Amazon) were converted to cropland. Perhaps most notably, soy production in Brazil grew from less than 20 million metric tons in 1990 to nearly 120 million metric tons in 2018.[129] During this time, more than 40 million hectares of the Amazon were deforested for agricultural purposes.[130] In the Cerrado, deforestation occurred at more than twice as fast as in the Amazon, contributing to a cumulative loss of more than 88 million hectares of native vegetation cover since 2000.[131]

With much of its land already titled in large parcels and with a transportation route mostly completed, the portions of the Cerrado located in Mato Grosso became a key site of neoliberal agricultural extractivism. As Brazil liberalized its economy and large-scale export-oriented agriculture expanded, idle but titled land in the state was rapidly converted into intensively farmed fields. Between 1985 and 2017, the amount of land in the state used for agricultural purposes increased from 37,835,651 to 54,850,819 hectares. At the same time, soy went from being planted on 2.1% to nearly 16% of land used for agricultural purposes, and the annual soy harvest increased from 1,610,530 to 29,281,387 tons a year. With over 80% of the agricultural land in Mato Grosso held by establishments of more than 400 hectares, this increase in production occurred on highly mechanized and technologically advanced farms.[132]

The replacement of Mato Grosso's native vegetation with cultivated crops changed more than the rural landscape. It also propelled the rapid growth of surrounding agricultural outposts. From 1991 to 2017, the estimated population of Mato Grosso increased from 2,026,069 to 3,344,544. During this time, four of the state's ten top-growing municipalities—Lucas do Rio Verde, Nova Mutum, Sorriso, and Sinop—were transformed into soy-producing hubs along BR 163 in the Teles Pires River basin.[133] These population increases were in part driven by farm owners and employees who commuted to the countryside for work.[134] The towns also became burgeoning business hubs for agricultural machine dealers, seed and chemical input suppliers, and storage and shipping facilities. In addition, a number of secondary services blossomed alongside those oriented largely to agricultural purposes. In such places, jobs generally paid better than in Brazil's urban coastal areas. While low-wage migrant workers were employed in both on- and off-farm work, median incomes and human development indices were higher and poverty rates were lower than national averages.[135] As a result, these agribusiness-focused municipalities expanded both vertically and horizontally, as residents settled in a diverse array of homes, from mansions with pools and towering apartment complexes in city centers and new suburbs, to multigenerational family homes, and makeshift shacks on the cities' edges.

The soyscapes of the Cerrado in Mato Grosso attracted not only people seeking to cash in on the global agricultural commodity boom, but also the opportunistic and ever hungry *Ae. aegypti* mosquito. As what was once a mosaic of grass and woodlands was turned into agricultural fields, the area's ecology was thrown off balance. In this known biological hotspot, the predators and competitors of *Ae. aegypti* had long kept the mosquito's population in check. Rapid agricultural development now reduced the area's native habitat by about half, significantly undercutting biodiversity and species abundance.[136] In Mato Grosso, new mosquito landscapes were thus developed and the insect's potential natural predators and competitors were displaced or disappeared.[137]

At the same time, the rapidly increasing populations of Brazil's agricultural outposts provided *Ae. aegypti* with ideal breeding grounds. The continual expansion of these towns created temporary breeding grounds for mosquitoes at building sites where the landscape had yet to be leveled and was thus more likely to be dotted with small patches of standing water. In addition, and perhaps more importantly, such towns were established as small rural outposts and originally offered limited public services. While centralized water service provision was present in most towns in Mato Grosso by the late 1990s, centralized sewage treatment and adequate stormwater drainage systems were not. Some towns began to establish sewage systems in the early 2000s, but others have only recently done so, and in 2016 many still lacked any such services. As a result, the residents of Mato Grosso's booming soy towns relied on septic tanks or had no formal means of sewage disposal. Within this context, the majority of residents in Mato Grosso's most rapidly growing towns had only semiadequate or inadequate access to sanitation services.[138]

Inadequate sanitation and drainage systems not only spawned places of mosquito reproduction but also boosted manifestation of *Ae. aegypti*-borne diseases. Since 2001, dengue outbreaks have regularly plagued Mato Grosso's rapidly growing agricultural outposts and with increasing intensity. In the early 2000s, dengue incidence rates in these places during outbreak years often exceeded those in Mato Grosso and Brazil as a whole, but only occasionally topped 5 cases per 1,000 residents. However, by 2009, dengue incidence rates in Lucas do Rio Verde, Nova Mutum, Sinop, and Sorriso all exceeded 10 cases per 1,000 residents. Over the following decade, dengue incidence rates in Lucas do Rio Verde have topped 10 cases per 1,000 residents three more times and in Nova Mutum and in Sorriso two more times. In Sinop, dengue rates have only twice dropped below 10 cases per 1,000 residents and have topped 30 cases per 1,000 residents three times. Such places also proved to be highly vulnerable to the Zika virus as it spread in 2015 and 2016. Most cases in Brazil occurred in larger metropolitan areas, with 33% of cases reported in municipalities with over 100,000 people. However, the Zika incidence rates in the towns with the fastest-growing populations—many of which grew alongside the soy industry—were among the highest in Mato Grosso. Indeed, nine of the ten most rapidly growing municipalities fell within the state's top twenty-five municipalities reporting the highest Zika incidence rates. While the earliest outbreaks of Zika were thus largely concentrated in Brazil's urban coastal centers, by the end of 2017 the interior state of Mato Grosso had the highest incidence rate and the third highest number of cases of the disease in the country.[139] And in all these sites of dengue and Zika in Mato Grosso, less than half of the population had access to adequate sanitation services.[140]

## FINANCIAL CONQUISTADORS

To Joshua Sokol, the "worst animal in the world" is the *Ae. aegypti* mosquito.[141] However, its ability to achieve "world domination" was far from its own doing. The mosquito was brought to Brazil by White European colonizers seeking to enrich themselves through the slave and sugar trades. As with Michael Farris's purported "northern invasion" of Virginia by "damn Yankee" ticks, such colonizers created the ecological conditions in Brazil for *Ae. aegypti* and the diseases it carries to thrive. Once established, the mosquito and its associated diseases wreaked havoc upon Brazil's coastal metropolises for centuries. However, it was not until the end of the twentieth century that *Ae. aegypti* began to plague the Brazilian interior.

As global financial flows shifted and the basis of Brazil's development model became neoliberal agricultural extractivism, agricultural fields became favored over the trees and grasses of the Cerrado. The landscape was turned into large monocrop farms, and a burgeoning array of agricultural outposts rapidly grew. The area's preagribusiness ecological balance was thereby lost. Unable or unwilling to invest in adequate infrastructure, such places became breeding grounds for *Ae. aegypti* mosquitoes and sites of transmission of dengue fever and the Zika virus.

While the *Ae. aegypti* mosquito is an undeniably virulent pest, it is more of an opportunist than a conquistador. Over the past four hundred years, the mosquito has traveled mainly to places it has been taken, and in turn it has adapted to the environments where it was brought. Anthropomorphizing *Ae. aegypti* as something that can achieve "world domination" thus distracts us from seeing those truly seeking world domination and from the underlying causes of the diseases the mosquito carries. Indeed, the only animals consciously seeking world domination since

*Ae. aegypti* arrived in Brazil are humans. Historically, however, not all of these animals actively engage in activities of conquest; largely it has been only those driven by capitalist logic, including, for example, finance capitalists. Seeking to dominate humans and the environment as they seek to maximize shareholder value and create new financial products, financiers have radically transformed the landscapes of Brazil and created a new home for *Ae. aegypti* and the diseases it can carry.

# 3 All That Remains Is Man and Mosquito

On a sunny day in 2018, Kelly sat with an elderly man named Richard outside his home on the slopes of Mt. Elgon in the Bududa District of Uganda. Richard had grown Arabica coffee his entire life. The British introduced Arabica coffee to the area in the early twentieth century, and today it is the primary cash crop for farming households. While the mountainous hillsides make large-scale agricultural production difficult, the fertile volcanic soils and ample rainfall make the slopes of Mt. Elgon a good environment for growing a variety of crops.[1] What initially made Bududa and the Bugisu region of Mt. Elgon an ideal place for Arabica coffee production, however, was the existing social and ecological system. Arabica coffee thrived in the shady, forested, and biodiverse landscape. In addition, it could be planted among subsistence crops. Arabica coffee thus provided a way for local residents and both the colonial and postcolonial states to profit. In the long history of Arabica coffee, it has generally been recognized as the most important commodity produced in the region. Richard's coffee-growing enterprise was no exception. Well into his sixties, coffee continued to be his main source of income.

While Richard depended on coffee for his livelihood, his story stands in stark contrast to the popular perceptions of Arabica coffee as a boon to the

Mt. Elgon region. According to Richard, the money he got from coffee was not what it once was. Richard said that prices were better in the past and the coffee sector used to operate with greater transparency: "Farmers were given information about where the coffee was being sold, and the price . . . but that died away. Now it is very hard to know where and for what [price] the coffee is being sold." Richard believed the changes in the coffee sector had contributed to his struggles to secure adequate food and pay the school fees for his younger children. These were not the only problems he mentioned, though. He linked the changes to another menace: mosquitoes and malaria. "The one thing with coffee," he reported, "it has more mosquitos. They usually hide in the leaves." Even though he wore long pants and gum boots when he worked in his coffee plots, often he "still gets bites." Like many farmers in the region, Richard continued to grow coffee but understood firsthand the potential harm the mosquitoes living in his plots posed. Over the years he had lost two children, one to malaria. But why, after decades of cultivating Arabica coffee, was Richard making less money and being bitten by more mosquitoes?

Richard did not necessarily understand the ins and outs of the global coffee trade, but he knew that being a coffee farmer had been better in the past. His ability to earn a living from coffee had been affected by the changes that occurred in the coffee industry in Uganda and beyond. In the 1980s, global finance started to play a greater role in so-called global development. In addition, changes in the global coffee trade expanded the trade in coffee-based financial derivatives. In Uganda, the state's acceptance of conditional loans and aid allowed transnational firms to participate in and eventually monopolize coffee processing, purchasing, and export. At the same time, the increasing trade in coffee-based derivatives in international commodity exchanges resulted in higher levels of price volatility for coffee farmers. In response, Ugandan coffee growers were encouraged to offset such risk by adopting new cultivation practices aimed at maximizing production volume over quality and sustainability. On Mt. Elgon, this has meant clearing more land for coffee production and shifting away from the shade-grown and intercropping practices of the past. And in so doing, farmers such as Richard have been driven to alter the local environment in ways that make it more conducive to the presence of *Anopheles gambiae* mosquitoes and outbreaks of malaria.

## THE EARLY HISTORY OF *ANOPHELES* MOSQUITOES
## AND MALARIA IN UGANDA

The *Anopheles* mosquito is thought to have originated sometime in the Cretaceous period and gradually evolved into its current form in Africa.[2] *Anopheles* mosquitoes have been transmitting malaria to humans for thousands of years. Some studies suggest that the mosquitoes started biting humans around 10,000 years ago with the development of agriculture. Agriculture facilitated denser settlements and more sedentary lifestyles, making it easier for the mosquitoes to prey on humans than on animals.[3] However, scientists also note that the genetic development of sickle-shaped red blood cells in some people that allow them to survive certain strains of malaria demonstrates that the mosquito, the virus, and humans coevolved, perhaps over centuries.[4]

*Anopheles* mosquitos are the sole vectors of mammalian malaria. Hundreds of different species of *Anopheles* mosquitoes exist, but only seventy or so have been shown to be competent vectors that can transmit malaria to humans. The *Anopheles gambiae* is the main mosquito vector for malaria in Africa.[5] Malaria is caused by blood parasites in the genus *Plasmodium*. *Plasmodium* parasites rely on two hosts throughout their life cycle: a human (or another mammal) and a mosquito. *Plasmodium falciparum* is the dominant variety linked to malaria in Uganda and sub-Saharan Africa more broadly. While *P. falciparum* has been found in non-human mammals in some parts of the world, most cases of malaria in humans are thought to occur when *A. gambiae* carry the *P. falciparum* parasite from one human to another.

As with the proliferation of *Ixodes scapularis* and *Ae. aegypti*, the proliferation of the *A. gambiae* and the diseases it carries is as much a story of human-induced environmental change as it is of ecology and evolution. Research by environmental historians demonstrates that early Africans were consumers of nature but recognized its importance to their livelihoods and were thus careful caretakers of places in which they lived.[6] Natural resources were viewed as common property, and the conservation of wildlife and wild fauna was steeped in community-based rules, beliefs, and values. The forests were viewed as sources of fruits and grains, meat, medicines, and construction materials. Even when early African societies started securing more reliable and readily available food through the domestication of livestock, the forests were still an enduring source of a wide range of resources and food.[7]

Through carefully managed environmental practices, the precolonial landscape of East Africa was likely less amenable to *A. gambiae* and the spread of malaria. The vast forests were less fragmented and more ecologically diverse.[8] On the forest floor, sunlight was heavily filtered and temperatures remained cooler. In addition, the ground was more likely to be covered with a thick debris of fallen leaves, which more easily absorbed rainwater. In the places where water did accumulate, it was likely highly acidic. According to contemporary ecologists, shade, lower temperature, and highly acidic standing water can all deter the reproduction and development of *A. gambiae* larvae and offspring.[9] The more biodiverse precolonial forested environment in Uganda may also have tempered the presence of *A. gambiae*. On one hand, greater numbers of birds, frogs, and other species likely served as predators of mosquitoes and mosquito larvae. On the other, the wider variety of mosquito hosts available, such as monkeys and other mammals, may have provided *A. gambiae* with other food sources and in turn reduced the potential transmission of *P. falciparum* parasites to humans.[10]

Uganda officially became a British protectorate in 1894, and the colonizer began to alter the landscape in ways that made it more conducive to the presence of *A. gambiae*. The British introduced and sought to grow and export American upland cotton.[11] In addition, the colonizer worked to expand coffee production. While robusta coffee (*Coffea canephora robusta*) was native to the region and grew naturally in some parts of the colony, Arabica coffee (*Coffea arabica*) was introduced to the Ugandan highlands in 1912.[12] Despite the early establishment of some plantation agriculture in the country, most cotton and coffee production occurred on smallholder subsistence farms. The British ruled out settler forms of colonization in Uganda early on and granted land concessions to local populations shortly after it assumed control of the territory. With a nine-month rainy season and subtropical climate, Ugandan farmers could plant and harvest two crops a year. Many could grow nearly enough food for subsistence in one cycle. If they did, more cotton could be planted in the second cycle.[13] With access to land on which to subsist, however, local populations had no reason to engage in wage labor on plantations. In addition, land registration for purportedly unused land was known to be slow, particularly in comparison to the neighboring British colony of Kenya. The British thus had difficulties recruiting foreign nationals to come to the country to engage in large-scale agricultural production. As a result, the colonizer coerced local populations to plant and sell cash crops but developed and controlled the secondary industries necessary to profit from agricultural exports.

After 1929, such efforts were bolstered through the Colonial Development Act (CDA). The CDA allocated some of the taxes that the British gleaned from their colonies to "the purpose of aiding and developing agriculture and industry . . . and thereby promoting commerce with or industry in the United Kingdom."[14] With CDA funds earmarked for nonrecurring capital expenditures, the money going to Uganda largely went toward the development and maintenance of transportation and communication infrastructure.[15] Roads opened more territory to cash crop agricultural production for the British and decreased the costs of trade. As noted in the United Kingdom's 1931 East Africa Commission report, "There can be no doubt that road development in Uganda has been largely responsible for the great increase in the growth of cotton . . . money spent on roads is doubly valuable, as it not only encourages the cultivation of economic crops but, almost as important, encourages the circulation of traders and natives, and increases the natives' wants."[16]

The CDA was revised as the Colonial Development and Welfare Act (CDWA) in 1940 and again in 1945. Improved transportation and communication infrastructure did boost the production of cotton and coffee in Uganda, but such increases were the result of cultivating additional acres of land. Per acre yields largely remained stagnant.[17] To the British authorities, the unchanged productivity was attributed to a lack of understanding among Ugandan farmers about how to best grow cash crops. The CDWA thus earmarked funds for both capital and recurring expenditures. While the allocation of CDWA funds was derailed with the advent of World War II, education—especially that promoted through agricultural extension— received the largest share of funds in the postwar era.[18] In Uganda, funds went not only to technical agricultural training but also to agricultural storage and processing facilities.[19]

After World War II, the British had an added incentive to increase agricultural yields. The war had substantially altered Britain's position in the world. As the colonial secretary admitted, "In the six years of the war, the UK changed from one of the major creditor countries of the world to the world's principal debtor nation."[20] Heavily indebted and short of dollars as a medium of foreign exchange, the British thus sought to increase the export of dollar-earning commodities. While cotton and coffee grown in Uganda had a market in Britain, they could also be sold in the United States. Increasing the production of both cash crops could thus provide the British with a way to obtain dollars that could go toward their debt.[21]

At the same time, the colonial authorities in Uganda faced growing threats of unrest. Cotton and coffee farmers began to question and challenge the oligopolistic power granted to a handful of processors and traders. To increase their bargaining position, Ugandan cotton and coffee farmers had already begun to organize informal cooperatives in the 1920s and 1930s, but such cooperatives were not legally recognized. While the British had helped develop cooperatives in some of its colonies outside Africa for decades, in Uganda, cotton and coffee processors and traders saw cooperatives as a threat to their own power and profit-making potential.[22] By 1945, however, the British Colonial Office began to promote cooperatives more actively in all of its colonies as a potential way to simultaneously enhance cash crop production, quell unrest, and increase the UK's supply of dollars.

Drawing on their experiences in India, the British believed cooperatives could be used to educate rural communities about the ways of commerce and thus to better incorporate them into capitalist circuits of agricultural production.[23] In particular, the British saw cooperatives as a way to educate farmers on modern agriculture and the inner workings of debt and finance. Through cooperatives, British "experts" believed they could not only train farmers how to improve production but also encourage them to collectively invest in and deploy modern agricultural techniques. Cooperatives could also help farmers internalize money-lending practices, thus helping British authorities to shift responsibility for supporting high-risk borrowers from themselves and British banks. The Colonial Office's cooperative ordinance was adopted in Uganda in 1946. However, the rules governing its implementation were clearly written to favor the interests of foreign—mostly British—agricultural processors and traders. As a result, few Ugandan farmer organizations registered as formal cooperatives. Amid continuing threats of unrest from cotton and coffee farmers, the British modified their rules regarding cooperatives in Uganda in 1952. The new rules both made it legal for cooperatives to have their own cotton- and coffee-processing facilities and provided financing for them to do so.[24]

Under British rule and during the first half of the twentieth century, cotton and coffee production dramatically increased in Uganda. While cotton production grew slowly after its initial introduction, by 1935 over 50,000 tons of cotton was being cultivated on approximately 1.5 million acres annually.[25] In the coffee sector, cultivation increased from less than 5,000 tons on approximately 58,000 acres in the early 1910s to more than 60,000 tons on more than 300,000 acres by the 1950s.[26] Cotton production outpaced coffee production from the time of its introduction through the 1940s. Coffee, however, was worth significantly more per unit of output and had substantially lower labor requirements and fewer material inputs; thus it gradually replaced cotton in areas where both could be grown. Coffee could also be interplanted with food crops more easily, affording farmers greater autonomy by allowing for the greater production of subsistence crops. Whereas cotton and coffee together on average accounted for over 80% of Uganda's total export value for much of the first half of the twentieth century, coffee surpassed cotton as Uganda's most valuable export commodity in 1955.[27]

The expansion of cotton and coffee production not only increased British profits from the colony but also resulted in the creation of environments more hospitable to the presence of the *A. gambiae* mosquito and the spread of the *P. falciparum* parasite. Replacing forests with agricultural fields—in particular for cotton cultivation—resulted in the creation of places with greater amounts of standing water for mosquitos to lay their eggs in and fewer mosquito predators. And the more time people spent in such fields planting and harvesting crops, the more likely they were to be bit by *A. gambiae* and contract malaria. While the British deployed a number of techniques to control *A. gambiae* populations, malaria rates steadily increased in colonial times. In 1911, the British documented 1.86 cases of malaria per 1,000 people. By the late 1930s, malaria cases topped 20 per 1,000 people, and by the late 1950s, malaria cases averaged more than 30 cases per 1,000 people.[28]

## THE SPILLOVER: FINANCE, LANDSCAPE CHANGE, AND THE REMAKING OF MALARIAL ENVIRONMENTS

In 1962, Uganda achieved independence from British rule. The colonial legacy of cash crop cotton and coffee production left Uganda dependent on its agricultural sector for foreign exchange. To secure higher returns to Ugandan farmers for their crops and somewhat insulate them from global market fluctuations, the newly independent Ugandan state worked closely with its agricultural cooperatives and fostered the creation of state-run marketing boards. A military takeover of the Ugandan state in 1971, however, weakened the cooperative structure and made the centralized state-marketing strategies less effective. In turn, cotton and coffee production and exports radically declined. When Uganda returned to democratic rule in 1979, it was forced to seek out external financial help and found that the system of global agricultural commodity trade had changed. In an increasingly financialized world, Uganda's agricultural sector had to conform to the conditionalities of IMF and World Bank loans, and the livelihoods of the country's farmers became subject to the whims of agricultural commodity exchange traders speculatively buying and selling new financial assets—in particular coffee derivatives. And as Uganda's farmers heeded the advice trickling down from the advisers of international lending agencies and attempted to navigate increasing price volatility, they changed their farming practices in ways that enhanced the presence of *A. gambiae* and in turn malaria.

In the years immediately following independence, coffee remained Uganda's most profitable export. Agricultural policy in Uganda was thus heavily shaped by the International Coffee Agreement (ICA). In the late 1950s, an oversupply of coffee in the global marketplace led to steadily declining prices, leading coffee-producing countries to sign an agreement in 1959 to limit their exports to agreed-on quotas. Without a means of enforcement, the agreement was largely ineffective. However, a new agreement was reached in 1962. With a stamp of approval from the

United States, the agreement tasked consuming countries with enforcing export quotas by collectively monitoring and reporting where their coffee imports came from.[29] While the ICA was intended primarily to help stabilize global coffee prices, it also slowed the expansion of coffee production in countries such as Uganda. In the decade prior to the 1962 ICA, Uganda's coffee production had grown by approximately 14%. In the decade after, it grew by less than 5%.[30] The ICA thus had an unintended impact: it led to a greater focus on enhancing the quality of the coffee produced than on boosting the quantity and allowed the newly independent Ugandan state to more fully support its agricultural cooperatives.

To improve the quality of agricultural commodities grown in Uganda, cooperatives worked with state marketing boards to educate farmers and furnish them with necessary inputs such as shovels, wheelbarrows, and seeds. These efforts would help farmers learn how to better prepare the land for cultivation. With coffee in particular, cooperatives sought to teach members how to plant and care for coffee plants using shade-grown techniques that both protected the soil from erosion and improved the quality of the coffee produced. In addition, cooperatives helped farmers purchase pulping machines so they could put the coffee beans they produced through the first stage of processing and thereby get higher prices for them.[31]

Cooperatives in postindependence Uganda also sought to capture greater profits for their members by continuing to follow through on their colonial-era push to play a prominent role in the processing and sale of agricultural commodities. By the mid-1960s, cooperatives owned 42 of 115 cotton ginneries as well as many coffee-processing facilities. In addition, 80% of trading licenses were held by local Ugandans.[32] As cooperatives expanded their roles in the processing, storage, and trade of cotton and coffee, more Ugandan farmers joined and sold their crops through their local cooperative societies.[33] In cooperatives, farmers worked directly with locally appointed officials to sell their crops. After harvest, members would bring their cotton and coffee to the cooperative facility, where it was weighed and graded. A small payment was immediately given to the farmer at a fixed rate. After some basic processing in the gin or pulping machine, the cotton and coffee were transported to state marketing board warehouses. There the crops' quality and weight were further assessed and farmers received a second payment. The state marketing boards then bulked together similar grades and classes of cotton and coffee and sold the commodities to foreign importers. Any profits made above what farmers had already been paid were later distributed through the cooperatives based proportionately on the quantities and qualities the cooperatives delivered and the average price per grade the state marketing board received through its bulked sales over the year.[34] Farmers were thus paid the same price per grade irrespective of when they had delivered their crops to the cooperative and regardless of when their particular grade of cotton or coffee was actually sold on the international market. The risk of selling a crop at the wrong time was thus removed from the farmer and born by the cooperatives and state marketing boards.[35]

In strengthening its agricultural sectors—in particular that for coffee—through the cooperative system, Uganda experienced robust economic growth in its first decade of independence. With increased investments in infrastructure and education, malaria rates stabilized and even dropped in some parts of the country.[36] Declining malaria incidence was also facilitated at the time by global efforts to combat the disease. Throughout the 1960s, the postcolonial Ugandan state worked with the World Health Organization on a number of eradication projects. It used DDT and malathion, as well as indoor residual sprays, to control mosquitoes and distributed chloroquine tablets and chloroquine-medicated salt to malaria-burdened areas in attempts to ward off the disease in humans.[37]

After global demand for primary commodities slowed in the late 1960s, political and economic instability gripped Uganda in the 1970s. The military government that seized power in the 1971 coup appointed officials with little sectoral knowledge to positions of power in cotton and coffee cooperatives and the state marketing boards. State support for the production of the two cash crops subsequently declined. Despite significant increases in the value of cotton and coffee in global markets, the prices farmers received for their crops from the state marketing boards stagnated.[38] In turn, the reported amount of cotton produced decreased from 84,800 tons to in 1970 to 11,000 tons in 1978. The reported amount of coffee produced also decreased from around 221,000 tons in 1970 to 80,000 tons in 1978.[39] Record frost in Brazil—the world's largest producer of coffee—in 1973 and again in 1975 drove up the price of coffee and temporarily incentivized farmers to continue some production under the military regime. In addition, a number of Ugandan producers were able to smuggle their coffee into neighboring countries to receive better prices.[40] But many other farmers abandoned the production of cotton and coffee altogether and returned to subsistence agriculture for much of the decade. Although health services were also disrupted during this time, the reduction in coffee and cotton cultivation likely somewhat tempered the creation of new *A. gambiae* environments.[41]

After Uganda returned to democratic rule in 1979, the government sought to revitalize its agricultural export economy. However, such revitalization proved to be difficult. After nearly a decade under military rule, the cooperative and state marketing board system were in disarray, and Ugandan cotton and coffee production was lower than it had been in decades. On the global level, the oil shock of the late 1970s, a 50% drop in the price of coffee in the early 1980s, and the onset of the world recession further deteriorated Uganda's terms of trade.[42] The government's revitalization efforts thus came to depend on international finance and aid. While the past decade of political and economic instability in the country made most private banks and bilateral agencies hesitant to lend money to the country, the IMF and World Bank proved to be willing lenders.[43] As a result, Uganda's total foreign debt rose from approximately US$588 million in 1979, to US$2.6 billion in 1990, to US$3.53 billion in 2000.[44]

The IMF and World Bank conditioned their loans to Uganda on the adoption of an array of liberal economic policies. Their initial loans were tied to introducing floating currency exchange rates, changing the basis of import and excise duties to the value of goods and services, and eliminating price controls. For coffee and cotton producers, this at first led to a substantial increase in prices and a return to selling their crops in the formal export market. The price increases were largely offset, however, by a higher inflation rate.[45] In 1983, the World Bank entered into negotiations with the government of Uganda for the Agricultural Rehabilitation Project, a loan program that did provide funds to rehabilitate Uganda's crop-processing facilities, including twenty coffee hulleries. However, funds were also provided that would ultimately diminish the role of cooperatives and the state in the pricing, purchasing, and sale of agricultural commodities. This loan program entailed funding for consultants to examine the structure of the state's agricultural marketing boards to determine whether they should relinquish their control over the buying and selling of coffee and cotton and divest from operations not directly related to the trading of those commodities. In addition, the World Bank conditioned its loan on the creation of an Agricultural Policy Committee and Agricultural Secretariat that would, with advice from expatriate agricultural economists, make policy recommendations to the government based on the findings of the consultants hired to examine the cotton and coffee sectors.[46]

By the late 1980s, the Agricultural Secretariat was advising the government that changes to the state marketing board system were needed in order to provide farmers with greater market-based incentives to increase the production of agricultural commodities. Following this purported finding, the World Bank and government of Uganda negotiated another major loan in 1990, known as the Agricultural Sector Adjustment Credit (ASAC). According to the World Bank, ASAC "provided a substantial push to the general liberalization of the Ugandan economy."[47] Fulfilling the conditions of the loan, the government of Uganda ended the coffee-marketing board's state-sponsored monopoly over the purchase, sale, and trade of coffee. It did this by liquidating the coffee-marketing board's assets in 1991 and splitting it into two entities—the Ugandan Coffee Development Authority (UCDA) and the Coffee Marketing Board, Ltd. (CMBL). The Ugandan Coffee Development Authority was set up as an oversight organization that regulated the coffee sector by monitoring and promoting coffee production and exports. It was also tasked with testing export consignments for minimum quality standards. The Coffee Marketing Board, Ltd., structured as a public commercial marketing organization, was intended to operate in a competitive marketing system. To make this possible, the UCDA opened the coffee sector to private traders. By the mid-1990s, more than a hundred private coffee traders operated in Uganda, and the Coffee Marketing Board, Ltd.'s share of coffee marketing dwindled into irrelevance.[48] In 1994, the World Bank issued another loan to Uganda to help it liquidate the cotton marketing board and replace it with the Cotton Development Organization (CDO). The CDO performed many of the same functions in the cotton sector that the UCDA performed in the coffee sector.[49] With cotton now making up only a minor share of Uganda's exports, no commercial public marketing board was created to compete in the sector. Over time, private transnational corporations also bought up the coffee- and cotton-processing facilities previously controlled by cooperatives.[50]

With coffee now accounting for the vast majority of Uganda's export value, the country was also impacted by large-scale changes in the global coffee trade. The International Coffee Agreement (ICA) expired in 1989, when the United States withdrew its support.[51] The end of the ICA led to widescale consolidation in the coffee trade and greater financial speculation in the coffee derivatives markets. As the ICA quotas were lifted, the global market was flooded with cheap coffee, and many coffee-importing firms went bankrupt. Within just a few years, the five largest firms controlled 40% of coffee imports worldwide.[52] At the same time, although coffee futures had been traded on the New York Coffee Exchange since 1882, the ICA's quota system had basically eliminated the ability to profit from speculative trading in coffee derivatives. With the end of the ICA, price volatility increased in the global coffee trade. Increasingly, therefore, coffee futures were thus not only purchased by coffee importers to hedge against price changes from the time they acquired coffee beans to the time they sold it to roasters, but also by speculators seeking to profit merely by buying and selling the derivatives. Between 1985 and 1994, the volume of coffee traded on the financial futures market increased from five to fifteen times the volume of physically traded coffee.[53] Speculators with no intention of buying or selling coffee in its actual physical form thus came to dominate the coffee futures market. Overall, greater consolidation allowed for coffee-importing firms to have the upper hand in setting export prices, and financial speculation subjected smallholder coffee producers to greater price volatility.

The changes to Uganda's coffee and cotton sectors did at first increase the share of global market prices that went to agricultural producers. When selling their crops in the cooperative and state marketing board system, producers had typically earned less than 40% of the export price. In the newly privatized system, producers usually earned 60% of the export price, sometimes even more during boom years.[54] The elimination of the state marketing boards made farmers more vulnerable to seasonal and global market price fluctuations. In addition, costs that were previously internalized by cooperatives, such as those incurred for inputs and equipment, were increasingly absorbed by individual farmers. In the

1990s, frost in Brazil initially kept global coffee prices high. Cotton prices, though less variable, were also relatively high at the time.[55] While the prices producers could command in the mid-1990s thus exceeded those in the late 1980s, the flood of coffee into the global marketplace in the late 1990s resulted in lower global prices. Combined with the Uganda shilling's (UGX) devaluation required as a condition of World Bank and IMF loans, Ugandan farmers were making no more for their crops in the early 2000s than they had been prior to the elimination of the state marketing boards.[56] By 2007, global coffee prices had rebounded somewhat, then leveled off. However, Ugandan producers' annual earnings were largely dictated by global market fluctuations.

In many ways, the changes made in the 1990s returned Uganda's agricultural sector to its colonial form. In both times, Uganda's exports were dominated by one agricultural commodity. Whereas cotton made up most of Uganda's exports in the early 1900s, coffee solidified its dominance as the major export by the second half of the twentieth century. Despite efforts to revitalize the cotton sector, it never rebounded from its losses in the 1970s, and today it constitutes only a minor portion of Uganda's exports. In the absence of state marketing boards, private foreign companies largely took control of the purchasing, processing, and sale of the primary exports. While the changes in Uganda's coffee sector allowed hundreds of traders to enter the marketplace, a small number of companies owned by foreign agricultural trading firms came to control most coffee-processing and export activities. Since 2010, five primary companies have held at least a 50% share of the coffee export market. Such companies include Ugacof Ltd., owned by Swiss agricultural trading and finance firm SUCAFINA; Olam Ltd., an agricultural trading company originally based in London that moved to Singapore; Kyagalanyi Coffee Ltd., owned by the British agricultural trading firm ED&F Man; the Louis Dreyfus Company, a French agricultural trading firm based in the Netherlands; Kawacom Ltd., owned by Swiss-based agricultural trading firm ECOM Agroindustrial Corp., Ltd.; and Ideal Quality Commodities, a Ugandan-based trading firm.[57]

Not only did the form of Uganda's increasingly liberalized agricultural sector resemble its colonial past, but so did its impact on the landscape. The transformations in the sector led to changes in coffee-growing practices and a deterioration of quality standards. What came to matter above all were volume and volume-based profits. In response to the increased price volatility introduced by market liberalization, speculative trading of coffee futures, and increased competition, farmers planted more coffee in an attempt to shield themselves from the potential effects of a bad price or crop year. As private firms expanded their influence in the sector, the UCDA also pushed for larger coffee harvests and more hectares planted in coffee. Whereas the hectares of harvested coffee in Uganda remained relatively constant for much of the second half of the twentieth century, they began to steadily increase starting in the mid-2000s. In 2005, approximately 345,000 hectares of coffee were harvested. By 2019, Uganda producers harvested the highest number of hectares to date: 569,427.[58]

The landscape changes in Uganda that occurred with the expansion of coffee production have likely contributed to the persistently high rates of malaria in the country. The absence of state and international intervention in malaria control efforts in Uganda led to increasing rates in the 1980s and 1990s, but a renewed focus on the disease thereafter led to a decline in malaria rates starting in the early 2000s. In 2015, malaria incidence rates in Uganda had decreased to 253 cases per 1,000 people.[59] As increasing amounts of land were devoted to cultivating coffee, continually controlling *A. gambiae* populations and curbing the spread of malaria proved difficult. Such difficulties have been compounded by mounting pesticide resistance among *A. gambiae* and increasing drug resistance of *P. falciparum* to treatment.[60] In response, national malaria rates in Uganda have ticked upward. In 2021, Uganda's estimated malaria incidence rate had increased to 283 cases per 1,000 people.[61] While demand for coffee from foreign countries ensures that farmers at least make some money from the coffee they grow, it is not without costs. Growing coffee poses a number of risks. Not only is there increasing price volatility, but contemporary cultivation strategies are leading to greater vulnerability to malaria.

## MAN AND MOSQUITO IN THE COFFEE-GROWING DISTRICT OF BUDUDA

The Bududa District, located on the eastern slope of Mt. Elgon in the Bugisu region of Uganda, has a long history of resistance. Throughout the

1890s, the first decade of British rule, the people of the Bugisu region remained largely independent. The British tactic of co-opting local leaders to divide and rule proved ineffective with decentralized communities that practiced communal forms of production.[62] It was not until the British brought in arabica coffee trees from Ethiopia in the early 1900s—and required Indigenous farmers to cultivate them—that the Bugisu region became economically valuable to the colonizer. Today, at least 80% of households in Bududa are involved in growing or harvesting coffee.[63] However, the expansion of coffee cultivation under British colonialism and in the postwar era differed from the expansion of coffee cultivation in Bududa today. Whereas the British and postcolonial states recognized that the value of coffee grown on Mt. Elgon was linked to maintaining its forested environment, the contemporary Ugandan state and the transnational traders controlling the coffee industry have not. Forced to compete as individuals in the global marketplace, Bududa's coffee growers have increasingly expanded their production and altered the landscapes of Mt. Elgon. And in this altered environment, the rates of malaria have increased.

Mt. Elgon is a biologically diverse montane forest region that spans the border of Uganda and Kenya. The mountain serves as a major rainfall catchment area that provides critical water supplies to Uganda's lowland regions and eventually Lake Victoria. The soils covering this extinct volcano's slopes are highly fertile but delicate. With a rich soil nutrient profile and high levels of rainfall, the region is home to an array of unique and endemic plant species.[64] In addition, it has been a site of subsistence farming for centuries. The mountain's topography, however, has long deterred larger-scale agricultural production. Despite high levels of soil fertility and rainfall, the mountain's slope and soil composition make the landscape more prone to landslides when disturbed.[65] The ecology of Mt. Elgon placed the Bududa District and the broader Bugisu region on the geographic and economic periphery of the early British protectorate. While the British saw Mt. Elgon's temperate climate as potentially desirable for future settler occupation, the inability to grow cotton left it outside the colonizer's early colonial focus.[66]

Mt. Elgon became more fully incorporated into Uganda and the global capitalist system as imported arabica coffee trees became established. While the ecological conditions in the area were less than amenable to cultivating row crops, the forested, temperate, wet, and cool climate was ideal for arabica coffee trees. For the British, arabica coffee proved to be a viable export commodity that could be grown where other cash crops could not. In addition, it provided local communities with the means to pay the required colonial tribute. By the 1950s, farmers in the Mt. Elgon region grew two-thirds of Uganda's arabica coffee, the value of which surpassed that of all of Uganda's cotton exports.[67]

While the introduction of arabica coffee trees brought commodity production to the Bugisu region, it also in some ways protected the ecology of Mt. Elgon. Subsistence producers cleared small plots for agricultural production, but they often intercropped arabica coffee trees or planted them nearby in more forested areas. The early production of arabica coffee on Mt. Elgon was thus both dependent on maintaining the existing ecological conditions and was worked into the existing practices of subsistence agriculture. The British recognized that Mt. Elgon served as a crucial water supply for significant swaths of the Ugandan lowlands, and sought to preserve timber resources for potential future use; the colonial government therefore actively protected the slopes of Mt. Elgon from overexploitation. The British promoted conservation and afforestation under their initial forest policy, implemented in 1929. Later colonial policy even more explicitly recognized the role of forests in the "maintenance of climatic conditions for agriculture" and "the preservation of water supplies."[68] That said, colonial forest policy was implemented largely to ensure the productive use and ecological capacity of activities deemed beneficial to the British protectorate's income-earning potential. While the production of arabica coffee could be done sustainably within the existing food system, the British still turned large swaths of Mt. Elgon into a forest reserve, in turn depriving many subsistence farmers of their livelihoods when their land was declared protected.[69]

In the early postcolonial era, the Ugandan government continued to maintain the protected areas and conservation tactics employed by the British. On Mt. Elgon, the actions of the Bugisu Cooperative Union (BCU; also known as the Bugisu Cooperative Society) and Uganda's participation in the International Coffee Agreement in some ways aided such efforts. As elsewhere in Uganda, cooperatives and the ICA insulated Bugisu coffee producers from full-fledged market competition. In regulating the amount of coffee that could be sold in the international marketplace, the ICA kept global prices more constant. In addition, the buying and selling practices of coffee cooperatives protected farmers from seasonal price fluctuations. As one coffee producer noted in reflecting on what others had told him about the past:

> Yes, so the Bugisu Cooperative Society pays for the coffee, and these people come back to distribute the money to the different farmers from where they have brought their coffee. And you are given according to how much coffee you provided. And then there is a time he told us about bonuses, in that when the cooperative society, the Bugisu Cooperative Society, makes a profit, they also come back and share, and that is where they used to call a bonus.

Insulated from global and seasonal price variations, farmers on Mt. Elgon could focus on the quality of their coffee over the quantity and, in turn, could adopt more sustainable practices. As another coffee producer told Kelly, "They used to mix the coffee, mix it with crops and those shade trees." While shade-grown arabica coffee beans tend to grow slower in the rainforest habitat, they create a much richer coffee flavor. BCU officials thus emphasized the importance of planting or preserving shade trees and supplied cooperative members with seedlings of native trees when distributing the seedlings of coffee plants. In turn, the BCU and coffee producers strived not only to grow a quality product but also to protect the region's forests, watershed, and communities.

The BCU's ability to promote sustainable economic and ecological practices in the Bugisu region began to erode in the 1970s under the clientelism and dysfunction of the military dictatorship. Inexperienced state appointees were placed in BCU leadership positions. In addition, the state marketing board began to purchase coffee from the BCU on credit that was later repaid erratically or not at all. Furthermore, the military government started to push toward a single coffee price for the entire country. Historically, the BCU had sold its members' arabica coffee at substantially higher prices than the robusta coffee grown in the lowlands. In 1970, the state marketing board bought arabica coffee from Bugisu at four times the price of robusta coffee. This price differential accounted for the extra land and labor required in the production and processing of arabica coffee beans and was reflected in the global market price. By 1980, the price differential had been cut in half. All of these changes threatened the profitability of arabica coffee production and the ability of the BCU to insulate its members from market fluctuations. As a result, many coffee producers on Mt. Elgon shifted their focus toward subsistence farming in order to ensure their own survival.[70] Coffee production on Mt. Elgon slightly rebounded in the 1980s. However, production fluctuated in response to highly variable global prices.

As elsewhere in Uganda, transnational trading firms began to take over the operations of local cooperatives on Mt. Elgon in the 1990s and 2000s. In the Bududa District, the dominant buyers of coffee in the Mt. Elgon region were Kyagalanyi and Kawacom, owned by global-commodity trading groups ED&F Man and ECOM Agroindustrial Corp., Ltd.. Kyagalanyi established an especially strong presence in Bududa. By 2009, the firm had bought or rented most of the primary processing field stations in the district, nearly all of which had formerly been managed by the BCU. Kyagalanyi also took over space at the main BCU processing and exporting facility in the nearby urban center of Mbale.[71] With field stations located near coffee-growing households that are connected to larger processing and exporting facilities in Mbale, Kyagalanyi has expanded so that it exercises significant control of both the local economy and the entire Ugandan domestic arabica coffee value chain.

Seeking to expand the volume of exported coffee, Kyagalanyi and other transnational trading firms have worked with the UCDA to increase coffee production on Mt. Elgon. To do so, the transnational trading firms began to provide genetically modified high-yield and sun-tolerant strains of arabica coffee seedlings to local producers.[72] The UCDA also changed its arabica coffee–growing guidelines to account for the newly introduced strains. According to the UCDA handbook, "If the land is under forest, it must be cleared thoroughly," as native trees compete with coffee plants for nutrients and water. In addition, "they reduce photosynthetic activity and cause elongation of internodes both of which result in lower yields." The UCDA also discouraged intercropping, noting that such practices heighten competition for water, nutrients, and light and are also more labor-intensive. The UCDA handbook also advises, "Maize, potatoes, and cassava have high nutrient demands and should not be mixed with coffee" and that "beans also encourage pests."[73] While some farmers continued to intercrop in order to offset the risks of cash crop production, most reported that they kept plots "only for coffee." As one farmer noted about contemporary growing practices, "In the case of some plants like cassava, or beans, and if there are many bananas, they can compete a lot with the coffee plantation. So they have to cut down those crops so they can easily plant the coffee."

The changes to Uganda's coffee sector had an array of ecological effects on Mt. Elgon. The drive to plant more coffee and reduce the intercropping of subsistence crops has led to the felling of trees and the creation of more frontier zones. As one coffee producer on Mt. Elgon observed, "Currently, people are cutting down trees because they claim that the trees spoil the coffee plantations, but this was not how it was done before." Indeed, such methods stand in stark contrast to past practices that promoted shade-grown coffee production within the existing forest. As another coffee producer in Bududa explained:

> Nobody has any idea of protecting the land anymore. Therefore, if you went to the river here, any time of the day when it rains, you will see what? Mud, brown water. All that means that it is soil erosion from above. There has been a lot of intensive soil erosion. Therefore, the years that we have been harvesting so much coffee we should be doing well. But no, we have poor soil because we no longer have proper management. Back during the seventies and eighties, there was a lot of effort on soil protection. Protection of the forests. There were those from the [cooperative] societies that would come, make sure you had large trees in your plot, and mixed [the coffee] with *matoke* [bananas], other food crops. But not anymore.

While sun-grown arabica coffee might have better yields, quality and environmental sustainability are sacrificed.

The changes to Uganda's coffee sector also adversely affected the people who live on the slopes of Mt. Elgon. Community members describe Bududa as having a "pocket to mouth" culture, meaning that if people make money, they spend it immediately, often on basic necessities such as food. Decreasing amounts of intercropping may increase coffee yields, but it has also shrunk subsistence agriculture and in turn access to food. The demise of the BCU and other cooperatives have made such problems worse. The transnational trading firms that replaced them are most interested in maximizing profits. For firms such as Kyagalanyi, this means purchasing coffee in the rawest form for the lowest price. As one farmer recalled, "During those days, when BCU had the monopoly, all the coffee was through BCU, but currently, because of liberalization . . . Middlemen, they sell to middlemen. We have now a coffee factory, Kyagalanyi, with processing, so most of them prefer selling [the coffee] wet [and unprocessed]. Because they get fast, fast money." And with its field stations set up all around the Bududa District, Kyagalanyi has sought to capture the most value from the coffee it purchases by making it easy for cash-strapped coffee producers to immediately sell their harvested beans wet.

The push to increase production of coffee and the growing difficulty of earning a living from it in Bududa have also gradually changed the size of coffee farms and forms of labor used to harvest the crop. While coffee production in the region was historically performed by small producers who worked the land they owned, farmers in the community note that coffee production is becoming "more monopolized, by big farmers," and that "small farming could dissipate over the next generation, because the youth don't want to, or the land and small profits isn't enough [anymore] for them to be able to support their families." This shift toward larger coffee farms has in turn resulted in an increase in the use of wage labor in coffee production. While larger operations enjoy economies of scale when buying and selling coffee, harvesting the crop remains highly labor-intensive. The changes to the global coffee trade and the conditional loans Uganda has accepted over the past four decades have thus done what the country's British colonizers strived but failed to do: develop plantation-style coffee cultivation in Bududa and turn local farmers into wage laborers. In 2020, coffee laborers were paid little, usually making around only 5,000 UGX (less than $2) per day.

The altered landscapes and livelihoods on Mt. Elgon are well suited to increase not only the profits of transnational coffee trading firms but also the presence of *A. gambiae* and diseases they carry. Larger monocrop coffee farms tend to alter the environment in ways that proliferate mosquito habitats. In addition, low-paid workers with no land of their own on which to subsist are forced to work long, tedious hours in order to make ends meet. And as they labor harvesting coffee beans, they are more likely to be in the coffee fields at dusk and dawn and thus more likely to get bitten by *A. gambiae*. The links between land use changes and malaria were all noticed by coffee farmers in Bududa. Many perceive that there are more mosquitoes in their coffee plots and in the monocropped coffee fields than in other agricultural plots or in the remaining forested areas. In the words of one coffee farmer, "The one with coffee has more mosquitos. They usually hide in the leaves of the coffee." As another farmer elaborated, "The coffee garden is very wet and very fertile and it has a lot of plants. They breed from there a lot." As in northern Virginia and Mato Grosso, the less disturbed and more biodiverse forested environment of Mt. Elgon's past kept insect vectors and the diseases they carried somewhat in check. In contrast, extensive deforestation and land use change are known to create wetter areas with a mix of sun and shade—an ideal environment for *A. gambiae*. In addition, deforested and monocrop ecosystems are less biodiverse, with fewer *A. gambiae* predators and fewer mammalian hosts to dilute the transmission of *P. falciparum* parasites to humans, who then contract malaria.

## THE CONTINUAL BITES OF CONDITIONAL LOANS AND DERIVATIVES

Coffee farming in Bududa has certainly changed over the years. The acceptance of conditional loans from the IMF and World Bank, the consolidation of coffee trading firms, and the speculative trade in coffee derivatives have altered how coffee is cultivated, processed, and brought into the global marketplace. Cooperatives and a centralized state marketing board used to provide price stability and promote cultivation of higher-quality coffee beans through the use of more sustainable production methods. Today a handful of international agricultural trading firms work in tandem with the Ugandan Coffee Development Authority to offset the risks of selling coffee in the global marketplace by promoting the production of higher quantities of coffee in ways that are degrading the ecological landscape.

Reflecting on these changes, many farmers in Bududa routinely note that one used to be able to make "a good living off of coffee." They talk about once being able to educate their children and buy metal roofs for their houses, but "nowadays, there is a lot of competition for coffee, so the profits are very low." Rising competition and volatility, combined with pressure to earn money in an increasingly market-based economy, means that there is more and more coffee cultivation in Bududa, creating undue pressures on the land. As one farmer aptly stated, "People earn less from coffee. . . . Also, land is less. So much of the land in Bududa is already claimed, so if people are expanding coffee, where are they expanding it to? Most of the land that they use is up the hills. . . . What they do there is, they usually cut down the trees." These practices are leading not only to soil erosion and landslides but also greater incidence of malaria. As one local leader described it, the Bududa District was once a dense forested area, "with trees with great big branches, and snakes, different monkeys, baboons, and even forest elephants. But now, all that remains is man and the mosquito."

The drivers of landscape transformation and the resulting increases in the presence of mosquitoes and outbreaks of malaria are unlikely to subside anytime soon. The global coffee trade is only becoming more consolidated.[74] In addition, as this book goes to press, coffee derivatives are trading at near all-time highs. In 2021, the volume of coffee traded in financial derivatives markets nearly 25 times that actually exported in physical form.[75] With speculators driving the price of coffee on international commodity exchanges, volatility is likely to remain high. In addition, international agricultural commodity trading conglomerates that both buy and sell physical coffee and hedge its price in the futures markets will likely glean the greatest profits from the global coffee trade. As long as these conglomerates are without a care for the local environment, the conditions for the people living in Bududa and other coffee-growing regions in the world will likely remain the same.

Late one evening while walking on a path in Bududa, Kelly encountered a familiar face. An elderly woman who helped "a neighbor" pick coffee during the harvest season for between 3,000 and 5,000 UGX (US$0.77–US$1.29) per day, depending on how many kilos she picked, was making her way home after a long day. She mentioned to Kelly that her back and arms were sore from the work. "There were very many mosquitos," she reported. "Everywhere you go in the coffee garden, you will find mosquitos there, and they will bite you." About a week later, Kelly spotted the woman again. This time she was visiting a local drug shop, taking her hard-earned money to purchase a generic version of Coartem, the leading malaria drug. As soon as she was well enough, she was back to work, outside among the mosquitoes in her neighbor's large coffee plot. She did everything she could to earn "some little money" for the next time her or her children fell ill with malaria.

# 4 The Treatment of Fringe Benefits

In December 1963, the Studebaker Corporation closed the doors of its manufacturing plant in South Bend, Indiana, ending its production of automobiles and trucks in the United States. At the time, the liability of the company's pension plan exceeded its assets by at least $15 million. The following year, Studebaker informed its former workers that retirees and retirement-eligible employees over the age of sixty would receive their full pension. Other vested employees received a lump sum payment worth only 15% of the value of their pension. Unvested employees, including all workers under the age of forty, received nothing. While Studebaker was not the first company to default on its pension obligations, the United Auto Workers used the incident to draw attention to the need to insure and protect participants in private sector pension plans. The federal government did not immediately act, but on September 2, 1974, President Gerald Ford signed the Employee Retirement Income Security Act (ERISA).[1]

In the decade after Studebaker's demise and as the provisions of ERISA were gaining traction in political circles, a woman in Connecticut named Polly Murray came down with a seemingly undiagnosable illness. Suffering from intermittent rashes, swollen knees, stiff joints, and a sore throat,

Murray had sought out help from more than twenty-four doctors, but none had been able to determine the causes of her ailments and some had written her off as a person who "subconsciously wants to be sick."[2] After watching similar symptoms develop in her children and neighbors, she began to gather her own data. In the fall of 1975, she called the state health department to report what she had been finding. Around the same time, another mother from the area, Judith Mensch, also contacted the state health department and the CDC in trying to understand how her children and several others in the area all could have juvenile rheumatoid arthritis, a relatively rare autoimmune disease. Eventually the suffering was traced to a bacterial illness later named Lyme disease after the area in which the two women lived.[3]

The death of Studebaker and the emergence of Lyme disease would seem to have little in common. However, ERISA and the financialization of American society would draw the two events together. In particular, a provision in ERISA preempted states from regulating or taxing fringe benefits for which private employers held the financial risk.[4] While the provision was written into the law to prevent states from overriding federal protections for employee pensions and to give private employers a tax break, it ended up fundamentally altering health care in America. Most notably, ERISA incentivized and provided employers with a way to offer scaled-down health care plans or an array of plans at different price points with different coverage levels. In addition, the law provided managed care organizations (MCOs)—which frequently controlled their costs by limiting and denying certain health care services and treatments—with an inroad into health care provision, which they came to dominate in the United States. With scaled-down plans and a health care industry whose profits were increasingly based on decreasing its use, even those with health insurance were not always able to obtain the services and treatments they needed. In the case of Lyme disease, most people now receive the two- to four-week antibiotic treatment recommended by the Centers for Disease Control and Prevention (CDC) and recover. However, some people go un- or misdiagnosed and others have lingering symptoms. Classified as untreatable outliers, they are on occasion denied treatment for their ailments. ERISA thus helped to create a more profitable health insurance industry that generates excellent returns for shareholders, but, like the Studebaker employees left without a pension, some people who have or believe they have Lyme disease are left combatting and finding treatment on their own. In other words, the shift in the American health care system to focus on the health of Wall Street returns over the health of humans made it increasingly difficult for people to receive a proper diagnosis of and treatment for Lyme disease.

## THE PATHOGENS OF LYME DISEASE

Lyme disease is the most reported vector-borne disease in the United States. More than 35,000 cases of the disease are reported annually to the CDC, but most cases go officially unreported. Studies based on insurance records suggest that approximately 476,000 Americans are annually diagnosed and treated for Lyme disease.[5] Cases have been reported in every state. However, the disease has historically afflicted more people in the Northeast than in other regions, with some of the highest incidence rates occurring in Maine, Vermont, New Hampshire, Delaware, Pennsylvania, Rhode Island, Connecticut, New Jersey, and New York. More recently, the number of reported cases have increased in the mid-Atlantic and upper-midwestern states, in particular Maryland, Virginia, Wisconsin, and Minnesota.

The initial effects of Lyme disease on the human body are often nondescript. Early symptoms frequently include fever, headache, fatigue, and muscle and joint aches but, according to the CDC, 70% to 80% of infected people develop another telltale sign of the disease—an erythema migrans (EM) rash. The EM rash usually first appears at the site where the Lyme disease's vector—the blacklegged tick—bit its victim, expanding gradually to 12 or more inches and sometimes having a bull's-eye appearance. The rash can appear anywhere between three and thirty days after the tick initially latches on to the human body.[6] Given that blacklegged ticks feed on their hosts only for three to seven days, those afflicted by Lyme disease may never know they had a tick on them or show symptoms until weeks after they have removed the tick from their body. For those who do not develop an EM rash, are unaware of having been bitten by a tick, or do not seek treatment, Lyme disease can progress in the human body and result in the expression of an array of additional and more serious symptoms. Signs of untreated Lyme disease include severe headaches and neck stiffness, dizziness, shortness of breath, EM rashes away from the site of the tick bite, facial palsy, severe joint pain and swelling, heart palpitations, inflation of the brain and spinal cord, nerve pain, and numbness or tingling in the hands or feet.[7]

The discovery of Lyme disease in 1975 was not without tribulation. After repeatedly being misdiagnosed, documenting clusters of a similar illness, and reaching out to public health officials, Murray and Mensch were put in touch with Yale rheumatologist Alan Steere. Unlike previous doctors, Steere did not write Murray off with an imagined illness. Murray presented him with an array of informally gathered epidemiological data from neighbors and surrounding communities.[8] In response, Steere assembled a team composed of researchers from Yale and the Connecticut Department of Public Health to study the disease clusters. Surveying local parents, doctors, and school nurses, Steere's team confirmed Murray's findings.[9] Most notably, several of the afflicted had developed an EM rash that was later followed by joint pain. In addition, the patients tended to live near town but in more wooded ex- and suburban areas.[10] The following year the researchers linked the disease clusters to ticks after a team of Yale entomologists did a tick survey that demonstrated a greater abundance of *Ixodes scapularis* in places where cases were clustered.[11]

Research from the early twentieth century had linked EM rashes to *Ixodes* tick bites in Europe. The research noted that patients who had developed EM rashes on occasion experienced neurological symptoms. After patients were treated successfully with antibiotics in the mid-twentieth century, it was further hypothesized that the EM rashes were caused by a spirochete bacterium. The United States reported isolated cases of EM rashes (treated with antibiotics) in the early 1970s. In addition, a cluster of EM rashes on patients was reported at the Naval Submarine Medical Center in Groton, Connecticut, in the summer of 1975. Drawing on existing studies in Europe, naval doctors surmised that the disease was transmitted to humans through a tick bite and successfully treated the ailment with antibiotics.

Despite the existing research linking EM rashes to ticks bites, Steere and his colleagues believed they had identified a new disease. Because the earliest patients reported joint pain as a common ailment, Steere's team saw that symptom as the disease's distinguishing feature and initially referred to it as Lyme arthritis. Furthermore, as a rheumatologist, Steere focused with his team on the rheumatological ailments of the afflicted. At the same time, they dismissed earlier studies for lacking control groups and questioned the scientific legitimacy of claims that antibiotics were an effective treatment.[12] As Steere's team wrote, "The large variation in the natural course of the disease makes it difficult to evaluate whether the observed improvement (i.e., antibiotics) in the individual patient would have occurred anyway."[13] They therefore dismissed hypotheses that the disease was caused by a bacterium and focused on possible overlaps with other arthropod-borne viruses that result in joint pain. As physician and medical historian Robert Aronowitz writes, "The Yale rheumatologists' decision to make arthritis prominent in the case definition they used to collect the initial pool of cases followed the common epidemiological practice of constructing a case definition that is most likely to distinguish people who have the disease from those who do not. As in many investigations of case clusters, however, one necessarily ends up with a disease that fits one's preconceived definition, akin to a Texas bull's eye: after the bullet hits the barn, the bull's eye is drawn around it."[14] It was not until 1982, when Wilhelm Burgdorfer isolated the same spirochete—later named *Borrelia burgdorferi*—in both blacklegged ticks and Lyme patients, that the causative agent of the disease was definitively determined.

## THE PROPHETS AND PROFITS OF MANAGED CARE AND EVIDENCE-BASED MEDICINE

Lyme disease emerged in the United States amid a shift in the provision of health care. In the 1970s, both the federal government and employers providing health insurance began to seek out ways to lower the cost of care. Cost increases in the medical sector were rapidly accelerating, and premiums for employer-provided health insurance were rising. In response, companies began to shift increasing costs onto workers in the form of health insurance deductibles and higher health insurance premiums. In addition, some smaller firms stopped providing coverage, and many larger firms eliminated or increased the cost of dependent coverage.[15] Two of the proposed solutions were "self-insurance" and "managed care."

Self-insurance came about in 1974 as an artifact of the Employee Retirement Income Security Act. For much of the postwar era, employers had offered health insurance to their employees by purchasing plans through health insurance companies. Health insurance prices were based on a community rate that was largely the same for all people within a given geographic area. Through such plans, health insurance companies held the financial risk but distributed it across an entire community.[16] ERISA regulated fringe benefits and was largely focused on insuring pensions for employees of companies like Studebaker that were in financial trouble. Section 514 of the law affected health care benefits by preempting states from regulating or taxing "any employee benefit plan." After lawsuits involving section 514 moved through the courts, it was eventually decided that it pertained to the benefits for which employers held the financial risk. For health insurance, this ruling meant that employers could capitalize on the nuances of ERISA by "self-insuring" their employees and in turn pooling health risks internally within their own company or organization. In practice, self-insured entities used their own assets to fund health insurance plans but still hired outside companies to administer them.[17]

ERISA and the move toward self-insurance fundamentally changed how the health insurance industry operated and what was covered by employee health care plans. With ERISA exempting self-insurance plans from state regulation, employers and health insurance companies could offer health benefits with different forms and quality of coverage to different employees or employment sectors. The ability of firms to "self-insure" only their own employees also meant the insurance industry had to reconceptualize how it calculated risks. In doing so, health insurance companies moved from assessing community risk to assigning what are known as "experience ratings." Whereas community risk assessments made insurance rates the same for everyone living in a defined area, experience ratings attempted to base insurance rates on the expected actuarial risks of individuals or groups of individuals. For self-insured employers, this meant the expected actuarial risk of their own workers. While calculating actuarial risk initially proved difficult, the move toward experience ratings gave birth to more sophisticated forms of medical underwriting that allowed insurers to identify and limit coverage for higher-risk applicants.[18]

In practice, the changes allowed some employers to potentially pay less for health insurance while simultaneously allowing the health insurance industry to bolster its profits. Larger employers could better pool their risks internally. In addition, employers with less risky forms of work could obtain better experience ratings. Both sets of employers could in turn secure lower health insurance rates for their employees. Smaller employers, though, faced higher costs, and people with higher-risk occupations—such as mining or logging—basically became impossible to insure. Insurers also sought to lower costs and boost profits by excluding individual high-risk employees in larger companies or riskier professions by limiting or denying coverage for things such as preexisting conditions.[19]

In addition excluding high-risk individuals from the health insurance marketplace, another prospect for lowering costs in the health care industry began to gain traction: managed care. Created under the auspices of economic efficiency, promoters of managed care sought to vertically integrate the activities of health care finance and insurance, physicians, and service delivery facilities, such as hospitals. Historically, the health care industry had operated on a fee-for-service model, and health insurers would pay what physicians billed. In contrast, managed care sought to create a prepaid or lower price for health care services that were provided by an approved network of doctors and service delivery facilities.[20] The first managed care plans were formed in the 1960s as "health maintenance organizations" (HMOs). While small in number, HMOs operated at a lower cost than much of the health care industry. To promote HMOs and their cost-saving measures, Congress passed the Health Maintenance Organization Act in 1973. Initially, the legislation had little impact. However, as employers sought to further lower the costs paid for employee health insurance in the 1980s, the number of HMOs and other forms of managed care proliferated. By 1991, there were more than five hundred HMOs that provided services to close to 40 million people. In addition, preferred provider organizations (PPOs) and point-of-service (POS) plans, two other forms of managed care, covered nearly 25% of American workers.[21] By 2020, over 70% of workers in the United States with employee-sponsored health plans were enrolled in an HMO, PPO, or POS plan.[22]

Managed care organizations employed a number of strategies in attempts to achieve greater economic efficiency. First, most MCOs sought to provide fewer services at lower costs by prioritizing the use of primary care physicians, preventive medicine, and outpatient treatment.[23] Second, many HMOs operated as insurers with their own doctors and facilities. With all major points of health care in-house, administrative overhead could technically be reduced.[24] Third, for managed care organizations using a network of physicians, prices for services were fixed or negotiated, further lowering potential costs. And fourth, when patients required treatment outside that provided by their primary care physician, MCOs frequently required a "utilization review" that required patients to receive approval before seeing a specialist or being admitted to a hospital.[25]

While utilization reviews were ways to control the cost of health care, MCOs justified their use in science and best-practice guidelines. Congress had already sought to shape the health care industry along similar lines through the 1972 amendment to the Social Security Act. Confronted with rising costs for health care, the government wanted assurances that the services Medicare paid for were necessary and of quality. The amendment led to the creation of professional standards review organizations (PSROs), relabeled as peer review organizations (PROs) in 1983. The task of PSROs was to determine whether health services upheld professional standards, were medically necessary, and were provided in "the most economical but medically appropriate settings."[26] In utilization reviews, MCO representatives used criteria developed by PSROs and groups of doctors hired by private insurers to determine if a physician's treatment recommendations were necessary. By the mid-1990s, more than 50% of MCOs had developed their own criteria and practice guidelines.[27]

While many of the tenets of managed care fell out of favor in the late 1990s and early 2000s, "evidence-based medicine" carried forward the ideas underpinning utilization reviews, in particular the use of science to inform best-practice guidelines. However, whereas MCOs emerged from the health care industry, evidence-based medicine emerged out of academic medical institutions. Proponents of evidence-based medicine defined it as "the conscientious, explicit, and judicious use of current best evidence in making decisions about the individual care of patients."[28] The advocates of evidence-based medicine asserted that a high degree of variation existed in the treatment of the same medical conditions and that some medical techniques were unnecessary, overused, and supported with questionable evidence. Echoing the concerns and beliefs of MCOs, they claimed the use of such medical techniques contributed to spiraling health care costs and resulted in potentially suboptimal health outcomes. To solve such problems, proponents of evidence-based medicine moved beyond PSROs to promote the use of randomized control trials and meta-analyses to develop research summaries and best-practice guidelines for clinicians.[29]

The benefits of evidence-based medicine were touted by an array of actors. Using medical treatments supposedly proven effective through randomized control trials, doctors were able to more effectively wield the tools of science to back up their decision-making processes.[30] Drawing on more established guidelines, evidence-based medicine also gave doctors road-maps with which to better explain medical procedures and diagnoses. Such information afforded doctors the ability to reveal to patients the expert knowledge behind medical recommendations.[31] But, perhaps most relevant, evidence-based medicine was also roundly supported by the insurance industry. While the original promoters of evidence-based medicine saw it as a way to prescribe the best health care regardless of cost, those paying for and profiting from health care saw evidence-based medicine as a way to better realize and control the costs of various treatments.[32]

In the United States, evidence-based medicine began to gain traction in the 1990s. While professional medical associations had long issued guidelines for doctors, most guidelines historically focused on training, certification, and licensure. In addition, peer review organizations focused only on cost and quality improvement for Medicare and Medicaid. In November 1989, Congress thus created the Agency for Health Care Policy and Research (AHCPR).[33] The AHCPR was tasked with arranging the development, periodic review, and update of both "clinically relevant guidelines" for use by "physicians, educators, and health care practitioners" and "standards of quality, performance measures, and medical review criteria."[34] While evidence-based medicine did not explicitly guide the AHCPR's initial decision-making process, it was incorporated into the agency's mission shortly after its creation. Between 1992 and 1996, the AHCPR released nineteen evidence-based clinical practice guidelines.

In 1997, the AHCPR also established "evidence-based practice centers," tasked with creating "systemic reports, research protocols, and research reports" to aid in the production of new clinical practice guidelines and the review of existing ones.[35] In 2010, the Patient Protection and Affordable Care Act further codified evidence-based medicine in the United States by requiring insurers to cover preventive services with a known, evidence-based net benefit.[36]

Despite its advocates and adoption in law, evidence-based medicine has had its fair share of critics. Using randomized control trials and meta-analyses to construct guidelines and best practices, evidence-based medicine shows the "comparative efficacy of treatment for an average randomized patient."[37] Guidelines derived through evidence-based medicine have thus been critiqued for failing to account for the individualized conditions underpinning or expressed by a disease. On one hand, creating standards and best practices based on the average randomized patient overlooks the different social and environmental conditions in which individuals live. On the other hand, focusing on the average randomized patient does not account for differences in the severity of patients' symptoms or in any potential comorbidities.[38] Such dynamics can influence not only the causes of a disease but also the effectiveness of different treatments. While the standards created through evidence-based medicine work for many, they can direct practitioners to see past other potential factors influencing a disease. In addition, and for those who do not fit the profile of the average randomized patient, the standardized treatment may not work. In such contexts, the patient can become labeled as potentially untreatable. A consequence of institutionalizing evidence-based medicine as the gold standard has thus been the creation of patient outliers.

Critics have also identified evidence-based medicine in the United States as a tool for the private health care sector to increase its profits. While federal policies require private insurers to cover some evidence-based services, they also make it easier for private insurers to deny coverage for other services that may be beneficial to some patients.[39] In addition, in the era of welfare state retrenchment, those best able to utilize the gold standards of evidence-based medicine techniques—randomized control trials and meta-analyses—have been in, affiliated with, or funded by the private health care sector. As John Ioannidis, professor of medicine at Stanford University and an early proponent of evidence-based medicine, has written, evidence-based medicine has been "hijacked to serve agendas different from what it originally aimed for. Influential randomized trials are largely done by and for the benefit of the industry. Meta-analyses and guidelines have become a factory, mostly also serving vested interests."[40]

## MANAGING THE EVIDENCE OF LYME DISEASE

The evidence-based guidelines informing doctors and managed care organizations about how to best diagnose and treat Lyme disease were developed by the Infectious Disease Society of America (IDSA). A professional organization composed of physicians, scientists, and public health experts, the IDSA identifies its top strategic priority as advancing its "role as a preeminent source of information and knowledge." To do so, the society seeks to "optimize the development, dissemination, and adoption of timely and relevant infectious disease guidance and guidelines to improve the outcomes of clinical care."[41] The IDSA has developed guidelines for close to a hundred infectious diseases. The IDSA guidelines for Lyme disease were created in collaboration with the American Academy of Neurology and the American Academy of Rheumatology. They were first released in 2000 and most recently updated in 2020.

The IDSA guidelines for the diagnosis of Lyme disease rest in testing patients' antibodies. The causative agent of the disease, *B. burgdorferi*, is often difficult to find in the human body and culture in a lab. To test for most bacteria, doctors seek to isolate a sample from a person's bodily fluids. However, *B. burgdorferi* tends not to stay in the bloodstream or spinal fluid for long, even when it remains present in body tissues.[42] Although the spirochete has been cultured from human blood and spinal fluid, the rate of successfully doing so has historically been highly variable, even among patients displaying visible symptoms that frequently demonstrate the presence of the bacteria, such as an EM rash. Taking this fact into account, the IDSA guidelines recommend clinical practitioners use a two-tiered antibody testing strategy that starts with an enzyme-linked immunosorbent assay (ELISA) and can be followed up with a western blot assay.[43] ELISAs detect bacterial antigens and are more sensitive than western blots. They are thus more likely to detect antibodies. However, ELISAs occasionally provide false positives. As a result, the IDSA recommends treating physicians use a western blot assay to confirm a positive ELISA. Western blot assays are more specific and separate out antibody proteins. In particular, western blot assays for Lyme detect both immunoglobulin M (IgM) and immunoglobulin G (IgG). IgM is present in larger quantities during the early stages of infection; IgG is present in larger quantities during later stages. A Lyme diagnosis is determined by the number of antigen bands that appear for each protein on a western blot assay. To be diagnosed as positive, 2 out of 3 IgM bands and 5 out of 10 IgG bands need to appear.[44]

The two-tiered testing regimen has been shown to have high levels of specificity, meaning that there are few false positives. However, the two-tiered testing regimen has lower levels of sensitivity, particularly during early stages of the infection. False negatives can thus occur if patients have yet to develop a strong antibody response.[45] While all medical tests have a level of error and both the IDSA and CDC recommend that doctors treat patients' clinical symptoms even if a Lyme test is negative, the latter has not always occurred. In the 1980s and 1990s, the disease was "new" and diagnostic guidelines were underdeveloped. In addition, both the expansion of Lyme geographically and its nonspecific symptomatology—particularly for patients who do not develop an EM rash—occasionally make it difficult for doctors to diagnose the disease. As one Lyme disease patient noted to Brent: "People can't find a doctor to treat them. I went through this from 1998 to 2006 or so, for eight years. . . . People don't know they have Lyme. Some go for months; some go for years." Doctors in places where the disease is new or less common are also more likely to misdiagnose or more stringently follow the IDSA guidelines. In patients without a clear EM rash, clinicians are more likely to wait for a blood test before confirming a diagnosis, lowering the likelihood of immediate treatment for patients who may have early-stage Lyme disease or test negative.[46] As a doctor who specialized in the treatment of Lyme disease but questioned the IDSA guidelines noted, "[If you're] an individual who is working in a run-of-a-mill clinic, not that familiar with the concepts, you're going to defer to the experts. The specificities of the two-tiered system are acceptable, . . . [but] the problem with Lyme is you have a 50% sensitivity, so you are going to miss people right off the bat."

The IDSA guidelines for the treatment of Lyme disease are based on a symptom-driven regimen of antibiotics. For patients with early-stage Lyme disease, a 14- to 21-day course of oral antibiotics is thought to rid the human body of the bacteria. For those with later-stage Lyme disease, the treatments vary. The IDSA recommends that people with neurological or cardiac symptoms be treated with a longer, 21-day treatment of antibiotics, occasionally distributed intravenously. For those suffering from severe joint pain and swelling, the IDSA recommends a 28-day course of antibiotics. In some instances, if symptoms persist, the IDSA recommends a two-week follow-up antibiotic treatment.[47] While the IDSA-recommended antibiotic treatment is effective for most people suffering from Lyme disease, some continue to experience symptoms such as fatigue, muscle aches, and joint pain. According to the IDSA, the prescribed antibiotic treatments eliminate *B. burgdorferi* from the human body. They thus label patients with lingering symptoms as suffering from "post treatment Lyme disease syndrome." Some Lyme disease patients and doctors believe *B. burgdorferi* can persist for longer periods of time in the human body, however, and they refer to the persistent symptoms as "chronic Lyme disease."[48]

In the evidence-based managed care system of private health care provision in the United States, the IDSA guidelines have sometimes restricted patients who initially tested negative for Lyme, were misdiagnosed, or had lingering symptoms from receiving certain types of treatment. As a doctor who did not always follow the IDSA guidelines told Brent: "For the IDSA, four to six [weeks of antibiotics] is all you need. Everything after is uninfectious or noninfectious and you don't need antibiotics after that. . . . They hide behind that dogma so they don't have to cover [the insurance payment]." A person who had suffered from late-stage Lyme and been denied coverage also noted, "They have their guidelines. You know: no more than four weeks of this antibiotic. You know, and so that's what insurance companies go by. And so, if you violate that in any way, shape, or form, they're just gonna throw your claim out." According to some doctors and patients, the IDSA's evidence-based guidelines thus more easily allow insurance companies to challenge or deny claims for longer-term use or nonoral forms of antibiotics, even when doctors believe such treatments may help their patients.

Some clinicians and patients also believe that diagnosing and treating patients with methods not in line with the IDSA guidelines may result in the investigation of a physician for medical malpractice. A patient who had lingering symptoms of Lyme disease and received treatment beyond the IDSA recommendations said: "Some doctors are willing to do it, but some don't want to get in trouble. 'Cause every doctor that I've been to has been investigated. And I've been to a lot. And I would say every good Lyme doctor has at some point been threatened with their license." A doctor who treated patients with longer regimens of antibiotics similarly noted, "There are doctors who want to treat Lyme. They're afraid of [losing] their licenses. They're afraid of insurance companies." While some physicians are willing to treat patients outside the IDSA guidelines, many of those who do fear retaliation for doing so.[49]

## THE INDIVIDUALIZATION OF RESPONSIBILITY FOR LYME DISEASE

From the beginning, the dismissal of the ailments of those afflicted by Lyme disease and the subsequent history of its so-called discovery in the United States placed the problems associated with the disease on and within the individual. After being reportedly misdiagnosed or told nothing is wrong with them, their kids, or their neighbors, Murray and Mensch had to gather their own evidence to demonstrate that an unusual disease cluster existed. In addition, Steere and his team's initial refusal to prescribe antibiotics led some patients in the early studies to seek help from other medical professionals. While the discovery of the causative agent of Lyme clarified how the disease could be diagnosed and treated, it did not remove the disease from the already highly individualized culture of American society. Evidence-based guidelines were developed to prevent, diagnose, and treat the disease. The recommended preventive measures, however, all involved individual action. In addition, while the recommended tests and treatments effectively identified and cured the disease in many people, outliers still existed. And, in an era of highly financialized health care, if one did not fit the profile of an average randomized patient, the burden of being properly diagnosed with or without Lyme, and in turn adequately treated, was further individualized.

Following the well-known trope that prevention is the best medicine, the IDSA and CDC recommend that individuals both protect themselves through chemical means and avoid places more conducive to the presence of blacklegged ticks. The IDSA and CDC advise that, during warmer months, when the ticks are more active, people wear permethrin-treated clothing or use insect repellents such as DEET or picaridin that are known to deter or kill ticks.[50] Although such neurotoxins, if used in diluted quantities, are not known to harm human health, both DEET and picaridin have been demonstrated to repel ticks, and permethrin is lethal to the parasites. For further protection, the CDC and local health departments also encourage homeowners to practice tick-safe landscaping.[51] People associate ticks and tick-borne diseases with forests and thus outdoor recreation areas, but many ticks are encountered in individuals' yards. While blacklegged ticks are unlikely to live in a well-trimmed lawn, they thrive near wooded edges, underneath shrubs, and in piles of leaves and damp wood storage areas. Taking this into account, the CDC recommends homeowners keep their lawns mowed, rake their leaves, and create mulch or gravel barriers between lawns and more wooded areas to restrict tick migration. In addition, the CDC notes that the use of acaricides can reduce the number of ticks in individuals' yards.

The most recent IDSA guidelines further individualize the prevention of Lyme disease by recommending antibiotic prophylaxis treatment. Whereas earlier guidelines advised against using antibiotics if a patient showed no symptoms or tested negative, the new IDSA guidelines recommend that people who have an "identified high-risk tick bite" be treated with a single dose of antibiotics. A bite from an identified *Ixodes* species in a highly endemic area that is attached for more than thirty-six hours is classified as high risk. However, identifying ticks and knowing how long they are attached is not always easy. *Ixodes* ticks are relatively small in their early life stages, making them easy to overlook when attached to the human body. The lay observer may also have difficulty distinguishing different types of ticks. Whether a person receives antibiotic prophylaxis treatment thus depends on their individual ability to find and properly identify an *Ixodes* tick on their body. According to the IDSA, "If a tick bite cannot be classified with a high level of certainty as a high-risk bite, a wait-and-watch approach is recommended."[52]

The individualization of responsibility for Lyme disease extends beyond prevention to obtaining the necessary care. People Brent spoke with in Lyme-inundated areas frequently expressed the need to find the right doctor in order to be diagnosed and treated. As a former director of one state health department said: "When somebody came in with an attached engorged deer tick, we would tell them go see your doctor. And on the side, I would tell some of them, if your doctor says no treatment, go see another doctor. Get a second opinion. . . . You want to do presumptive treatment if you can." Some patients expressed frustration over not knowing which tests to request after finding a tick bite and then how to read the results. A person diagnosed with Lyme disease years after initially experiencing symptoms told Brent: "As a patient you have to demand the testing. . . . You end up having to know more than you should know as a patient. You end up having to know what bands [of the western blot] corresponds to what. You know, you need a whole report. You need all your medical records, so that's . . . that's kind of tricky." The responsibility for finding and obtaining the desired care thus rests on individuals' knowledge of where to go and what to ask for.

In Lyme-inundated regions, the individualization of prevention and care left many feeling isolated. Like Paula Murray, who sought to draw attention to her own symptoms and the disease cluster in Lyme, Connecticut, in the 1970s, people are frequently led to believe that their suffering is of their own making. A public health professional in a heavily Lyme-affected area stated: "It absolutely gets pushed onto the individual and a lot of times it's in the context of, no one will help me, no one believes that there's something really wrong. Everyone thinks that I'm some—that my kid or myself is somehow making this up in their head." At the same time, patients who are misdiagnosed are led to believe they have an uncurable disease. In the words of a Lyme activist who also worked as a public school nurse and frequently saw kids with Lyme reported: "The child will come in, will have had a known tick bite. Maybe had a rash, maybe didn't. Maybe was tested for Lyme disease and found negative and then over the course of the next year develops fibromyalgia, chronic fatigue syndrome, POTS—postural orthostatic tachycardia syndrome. All these things with no known causes and no known cure, with a known tick bite in their history. And the parents are kind of like—know that it doesn't make sense. You know, they've got about five different doctors and none of them are looking at a bigger macro picture. . . . So it ends up falling in the individual's lap." Their ailments left un- or misdiagnosed, patients are led to believe there is nothing that can be done for them and are thus isolated in their search for a cure.

## ERISA AND COLLECTIVELY INSURING THE INDIVIDUALIZATION OF LYME DISEASE

Throughout the United States, patient advocacy groups have attempted to help those impacted by Lyme disease. On a basic level, such groups provide people a space where they can share their Lyme stories and exchange ways to navigate the disease and the health care system. Several of these groups have also organized to promote state legislation that would better educate people about Lyme disease and allow them to more easily receive varying forms of treatment. However, much of the legislation they have advocated further has codified the individualization of Lyme disease into law. In addition, such legal codification has had its limits.

In response to Lyme advocacy organizations, many states have passed laws to raise awareness about the disease and guard against potential errors in its diagnosis. Taking a more symbolic stance, some state governors and legislatures have formally recognized May as Lyme Disease Awareness Month. Such measures draw annual media attention and thus ideally serve as reminders to residents in tick-inundated areas to wear insect repellent, perform daily tick checks, and take other individual preventive measures. Some states have also passed measures to install signs at state-managed parks that warn visitors of the presence of blacklegged ticks and the potential threat of Lyme disease. Lyme advocacy groups in some states have further sought to challenge the legitimacy of the two-tiered testing system in diagnosing Lyme. Groups in Maine, New Hampshire, and Virginia have pushed for laws that require physicians to notify people who are tested for the disease about potential testing errors and the need for patients to seek out additional care if their symptoms continue.

Lyme advocacy groups have also more directly challenged the IDSA treatment guidelines for the disease and sought to assuage fears among physicians of repercussions for treating patients outside or beyond the established standards. Legislation has been debated in at least nine states to protect physicians who go against IDSA guidelines and prescribe long-term antibiotic treatment from being investigated by state medical boards and thus potentially losing their medical licenses. Furthermore, Lyme advocacy groups have targeted insurance companies' use of IDSA guidelines and evidence-based medicine to approve or deny coverage. Given that the IDSA does not support long-term antibiotic use for Lyme disease, the possibility for being denied coverage exists. Nearly a dozen states have considered bills that would require insurance companies to cover long-term Lyme disease treatments. Not all such state bills addressing the use of long-term antibiotics, physician protections, and insurance coverage requirements have been passed, but advocacy groups have succeeded in some states in getting such laws enacted and implemented.[53]

While Lyme advocacy groups have organized collective support for patients and pushed for legislation, most of their struggles have not countered the individualization of responsibility for the disease. Awareness and education about contracting Lyme disease still put the onus on people to take individual preventive measures to protect themselves. Highlighting the potential flaws in diagnosis has still left individuals with responsibility to seek out additional testing or find a doctor willing to look beyond the recommended guidelines, or both. And the push for protections surrounding the prescription of long-term antibiotics has still focused on treating Lyme disease in the individual body. Insurance coverage mandates could potentially offer a slightly broader, societal solution, but existing federal legislation pertaining to health care coverage has complicated such efforts. Most notably, ERISA preempts state-level legislation on health care for private companies that self-insure employees, which includes most people covered by managed care organizations. Furthermore, ERISA protects the organizations managing employers' self-insured plans from liability for denying coverage. Since the early 2000s, more than 50% of covered private sector workers in the United States have annually been enrolled in self-insured health plans.[54] With most of these plans following evidence-based guidelines such as those put forward by the IDSA, a segment of Americans could still be denied coverage for certain Lyme treatments or for any treatment if they went un- or misdiagnosed during the early stages of the disease.

## THE SHAREHOLDER VALUE OF LYME DISEASE

ERISA is often cast as a response to some corporations' inability to adequately fund worker pension plans. As such, its origins are attributed to the fiscally irresponsible behavior of a few bad companies and the scope of its impact is depicted as limited to employee fringe benefit plans. However, this framing overlooks the historical conditions in which ERISA emerged, the long-term impact the law has had on the American health care system, and how it affects the creation of medical knowledge and treatment of diseases such as Lyme. Indeed, the deeper origins of ERISA and its effects on Lyme disease diagnosis and treatment are linked to a restructuring of society that increasingly financialized and individualized everyday life.

While Studebaker defaulted on its employee pension obligations and was subsequently used to lobby for the creation of ERISA, the corporation's path to shutting its doors reflected these broader societal trends. In the late 1950s, the Studebaker Corporation was facing financial hardship and needed to refinance $54.7 million of its long-term debt. To receive the approval of creditors and shareholders, Studebaker sought out the help of the financier A. M. Sonnabend, whom *Business Week* called a "doctor of sick companies."[55] Sonnabend supposedly cured ailing companies by helping them acquire other profitable businesses slightly over their market value and then folding them into their business portfolios. The ailing company could then use federal tax loss loopholes to balance the books and again appear profitable. After Studebaker followed Sonnabend's advice, the strategy seemed to work. Shortly after Sonnabend was brought on board, he and the other Studebaker directors persuaded the company's creditors to forgive $21 million in debt by accepting $16.5 million in fifteen-year secured notes and $16.5 million in shares of preferred stock. In 1959, the price of Studebaker's common stock increased from $3 to nearly $29 a share. By September, all of the original creditors had sold their preferred-stock shares. It was estimated that the creditors could recuperate their losses if the preferred stock sold for $233 a share. Metropolitan Life Insurance—the last creditor holding Studebaker's preferred stock—sold each its 30,165 shares at $400.[56] The following year, however, the price of Studebaker common shares dropped below $10. In December 1960, Sonnabend resigned from his position with the company. Three years later, Studebaker ceased its production of cars. With no legal protections, the company's employees were left both without jobs and without pensions.

Just as Sonnabend's strategy was posited as a way to save Studebaker, so ERISA's passage was posited as a way to save worker pensions. But as financiers made a healthy profit off Studebaker at the expense of workers, ERISA would allow financiers to turn a healthy profit off an array of work-ers' fringe benefits. In a basic sense, the passage of ERISA helped usher in a shift away from defined-benefit to defined-contribution retirement plans. The latter shifted the risk of retirement plans away from corpora-tions onto individual employees. In addition, it fueled the creation of the financial services industry by making workers with retirement plans into shareholders. But ERISA also provided financiers with the opportunity to profit from another fringe benefit with the emergence of "self-insurance" in the provision of health care. Like the shift toward defined-contribution plans, the shift to self-insurance pushed more of the risk of health care onto individuals with the emergence of MCOs and more sophisticated forms of medical underwriting. ERISA thus inadvertently provided the grounds for further commodifying health care in the United States and, in turn, making the health insurance industry into a profitable investment vehicle for financiers and individual shareholders.

ERISA was never meant to affect the overall provision of health care or the individual treatment of Lyme disease. Prior to the law's passage, doc-tors depended more on clinical diagnoses, and their recommended treat-ments faced less scrutiny from health insurance companies. While the pre-ERISA provision of health care was far from perfect and the early unknowns surrounding Lyme disease would, in the law's absence, still have likely resulted in misdiagnoses, the structural conditions encourag-ing doctors or health insurance to deny treatment did not exist. However, with ERISA and the emergence of self-insurance, corporations began to offer less comprehensive health insurance plans to lower their costs, and health insurance companies began seeking ways to bolster profits for their shareholders by denying coverage. While medical science and evidence-based medicine were supposed to help rein in the costs and improve the quality of health care, their use by the health insurance industry has on occasion created patient outliers. For those potentially suffering from Lyme disease, some became outliers after initially going un- or misdiagnosed and later developing late-stage symptoms. Others became outliers by not fitting the standard Lyme patient profile, whether they lacked an EM rash or did not test positive through the two-tiered regimen. And, without an evidence-based diagnosis, many such patients have been denied care or coverage.

In the mid- to late 1970s and prior to the widespread expansion of self-insurance plans, the health insurance industry did not make exorbitant profits. From 1975 to 1979, most segments of the health insurance industry reported annual losses despite yearly premium increases averaging nearly 15%. Profit margins in the health insurance industry began to change in the 1980s. Increasingly, health providers and insurers shifted from being nonprofit to for-profit, publicly held companies.[57] More wedded to Wall Street, such companies sought to increase their profits and the returns for shareholders. While the initial returns in the transition to self-insured health insurance plans and for-profit health insurance companies rose only gradually, by the 2000s, health insurance companies had become some of the most profitable firms in the United States. Since 2010, the top five health insurance providers reported a combined average net income of more than $14.5 billion a year.[58] This industry shift has provided shareholders with exceptional returns. The share price of the largest provider of health insurance in the United States, United Health Group, increased from US$0.14 in 1984 to over $500 in 2022, providing annual returns of more than 20%.[59] The denial of coverage to some people afflicted by, or who believe they are suffering from, Lyme disease has contributed to such returns in shareholder value.

# 5 Mosquito Derivatives

On November 10, 2021, Oxitec Ltd. released a television commercial for a new product, designed to control "dengue mosquitoes." The commercial features young Oxitec employees walking individually through the streets of different urban areas in Brazil, carrying a box labeled with the product name, *Aedes do Bem*, meaning "Good *Aedes*." The first person who appears in the commercial touts the product as having been developed "in Brazil by innovative scientists and engineers." Another person advertises it as "safe" and "nontoxic," with its use reportedly having "no harmful impact on the environment." A third employee further notes that the product is a way to "protect our families, friends, and our neighbors, in balance with the environment."[1] Through Oxitec's discursive construction of its product, the company sought to quell the concerns that surrounded past mosquito control measures by promoting its product as a locally produced and environmentally friendly commodity that was good for all.

The product so advertised was a strain of Oxitec's genetically engineered mosquito known as the Friendly™ *Aedes aegypti*. Inside the box in the commercial were packets of eggs that would hatch genetically engineered mosquitoes. The Friendly™ *Ae. aegypti* was part of the company's trademarked suite of Friendly™ insects that contained what are known as

conditionally lethal or autocidal genes. Designed to reduce insect populations, the gene causes newborn offspring to suffer premature death. The
technology was first developed for the *Aedes aegypti* in the early 2000s.
Preliminary field trials in the Cayman Islands and Malaysia demonstrated
that the genetically engineered mosquito could successfully breed with
wild mosquitoes and showed some effectiveness at decreasing the local
populations. In February 2011, Oxitec sought to expand its field trials to
the Brazilian city of Juazeiro, Bahia. After releasing approximately 2,800
genetically modified mosquitoes a week for a year, scientists working on
the study reported an impressive 95% reduction of the local *Ae. aegypti*
population. The scientists also suggested that decreasing the mosquito's
population could potentially lower the transmission of dengue, but further studies were needed.[2]

The first published data on the seeming success of the Friendly™ *Ae.
aegypti* in an uncontrolled environment came out amid a mosquito-borne
disease outbreak in Brazil of the Zika virus. While the virus was likely circulating in the country in late 2014, it was not formally diagnosed until
the following year. The virus first appeared in a small cluster of patients in
the state of Maranhão. The patients expressed symptoms similar to those
associated with a mild form of dengue but tested negative for that better-
known disease. After two months of looking for another cause, doctors
determined that the patients were infected with the Zika virus. As Zika
continued to circulate in the country, its symptoms became more diverse.
On top of fevers, muscle and joint pain, rashes, and headaches, some
patients reported neurological complications such as Guillain-Barré syndrome. By early August, reports also began to surface that potentially
linked the disease to microcephaly, a condition in which babies are born
with smaller heads.[3]

As the incidence of Zika increased in the country and the results of the
Friendly™ *Ae. aegypti* study continued to circulate, the American company then named Intrexon Corporation entered into an agreement to purchase Oxitec for US$80 million in cash and US$80 million in stock.[4] At
the time, biotech billionaire Randal J. Kirk was chairman and chief execu-
tive officer of the company. His private investment firm, Third Security,
also owned a significant number of shares in Intrexon. Prior to expressing
interest in Oxitec, Intrexon had acquired an array of up-and-coming bio-

technology firms. Since going public in 2013, the price of Intrexon stock had fluctuated but hovered between US$20 to US$30 a share. In the weeks leading up to its purchase of Oxitec, the share price of Intrexon stock skyrocketed. On August 5, 2015, two days before Intrexon signed the purchasing agreement for Oxitec, it reached an all-time high of US$69.33 a share.[5]

The outbreak of Zika in Brazil undoubtedly helped raise the profile of Oxitec. However, the value of the Friendly™ mosquito is linked as much to shifts in global finance as to the emergence of a disease in a new location. In the United Kingdom, the founding of Oxitec was part of a broader campaign to defund and privatize higher education. Oxitec was started to allow private investors to fund and profit from the Friendly™ technology, developed by Oxford University scientists. While the Friendly™ mosquito was created in an Oxford laboratory, investors could not fully realize its value in the United Kingdom. The British Isles lacked a naturally sustainable population of *Ae. aegypti* and the diseases the mosquito carried were not present, so the mosquito had to be sold for use elsewhere. The *Ae. aegypti* and diseases such as dengue and Zika exist in many equatorial regions, but Brazil's changing political economy at the beginning of the twenty-first century offered a potentially profitable opportunity. As part of the push to liberalize the Brazilian economy, the country not only opened itself up to increasing flows of global finance but also decentralized decision-making power over how public funds should be used. The provision and means of health services, including the control of vector-borne diseases, was thus left to states and municipalities. Within this context and as international lenders intended, local entities could more independently choose how to control the *Ae. aegypti*, and Oxitec could sell its Friendly™ *Ae. aegypti* boxes in a purportedly open marketplace in one of the most populous countries in the world.

## THE PATHOGENS OF YELLOW FEVER, DENGUE, AND ZIKA

The pathogens transmitted by the bite of *Ae. aegypti* that cause yellow fever, dengue, and Zika have a number of common effects on the human body. Most people who become infected are asymptomatic or experience only mild symptoms. Symptoms that do develop are often nondescript and include fever, rash, body aches, or joint and muscle pain. For those who develop severe yellow fever or dengue, the diseases can cause vomiting, jaundice, shock, organ failure, and even death. While vaccines have made yellow fever less common, the virus still annually causes 30,000 deaths across the globe. In addition, dengue annually causes some 21,000 deaths worldwide.[6] The impacts of Zika are less severe, but newborns of women infected with the virus during pregnancy have an increased risk of being born with congenital microcephaly and other central nervous system malformations. During the initial outbreak of Zika in Brazil, more than 1,400 newborns were diagnosed with microcephaly linked to the virus.[7]

Since the early twentieth century, public health professionals have recognized that certain environmental conditions were more conducive to the proliferation of *Ae. aegypti* and the diseases it carried. In response, in some areas, fumigation was used to target adult mosquitoes. Primarily, though, policy makers sought to limit the presence of the mosquito by eliminating *Ae. aegypti* larvae through removal of standing pools of water. The installation of water and sewage systems in urban areas was thought to be one of the best ways to decrease the number of places *Ae. aegypti* could lay their eggs. However, because construction of these systems was unable to keep up with the rapid expansion of Brazil's urban areas, many people stored water in their homes. To keep such water supplies free of mosquito larvae, public health officials, in coordination with specialists from the Rockefeller Foundation, recommended dropping mosquito larvae–eating fish into water storage containers or pouring a layer of oil on top. To ensure that local residents adopted control efforts, public health officials also went door to door to check homes and properties for mosquitoes and potential egg-laying sites.[8]

Such methods began to change in the mid-twentieth century as doctors and scientists sought to internalize disease prevention in human and insect bodies. For the human body, an effective vaccine for yellow fever was developed in the 1930s, and its production and distribution began in Brazil in 1937. The following year, more than a million people were vaccinated.[9] Limiting the presence of *Ae. aegypti* still remained the primary strategy of disease prevention, however. With the greater availability and use of DDT in the postwar era, strategies to control *Ae. aegypti* shifted from targeting the mosquitoes' larvae to targeting the mosquitoes' bodies. Using DDT to control *Ae. aegypti*–borne disease required substantially less labor than previous methods. With its residual properties and effectiveness in killing mosquitoes over a long duration, DDT as a control measure required only one to three spraying sessions a year, instead of weekly inspections. As a result of the use of such strategies, in 1958 Brazil was declared free of *Ae. aegypti*.[10]

In the early to mid-twentieth century, Brazilian and international public health professionals working to address mosquito-borne disease in the country operated with a more holistic understanding of the human-environment linkages underpinning mosquito-borne disease. However, the shift to internalize disease prevention in human and insect bodies paved the way to refocusing public health efforts away from environmental causes. To some, the use of DDT to eradicate *Ae. aegypti* made the need to engage in larger-scale infrastructural projects to eliminate standing water less necessary. Public health campaigns also succeeded in vaccinating most people against yellow fever in the endemic interior regions of the country.[11] But, by the late 1960s, concerns about the environmental and human harms of DDT began to emerge. In addition, *Ae. aegypti* had begun to develop resistance to the chemical's effects. These developments contributed to the reemergence of *Ae. aegypti* in Brazil in 1967. While it was again eradicated, it reappeared in 1976.[12] A highly vaccinated population prevented the return of yellow fever, but by the 1980s dengue fever began to again plague Brazil.

## DECENTRALIZATION, HEALTH CARE, AND *AE. AEGYPTI* CONTROL

The Brazilian response to the reemergence of *Ae. aegypti* and the return of dengue was affected by temporary constraints on lending by private global finance, the shifting terrain of international public health, and the country's changing political dynamics. In the 1980s, Brazil's skyrocketing levels of debt made it difficult for the state to fund either larger-scale infrastructural projects or more labor-intensive monitoring and eradication programs.[13] While the World Health Organization had led the way in funding health programs across the globe in the postwar era, the World Bank began to enter the public health arena in the 1980s. As the World Bank gradually became one of the primary funders of health programs across the globe, it brought its mantra of decentralization into the health care sector. In Brazil, too, local social movements had, since the 1970s, called for the decentralization of health services to combat corruption and increase accountability.[14] In addition, some public health professionals saw decentralization as a way to move beyond clinical and epidemiological observations of disease, which had informed public health decision making in Brazil since the mid-twentieth century, and thus focus more on how the social and environmental conditions in which distinct populations live affect their health.[15] However, whereas local social movements saw decentralization as a key mechanism for achieving equitable and universal health care, the World Bank saw it as key to lowering costs and further opening the health care sector to private providers.

The World Bank explicitly sought to improve the economic efficiency of the health care system, decrease the cost of basic health services, and finance projects to control endemic disease. The first project it funded in Brazil, part of a broader development project for the northwest region, allocated US$13 million to address health problems in the area by bolstering agricultural and forestry programs. The funds were earmarked for the expansion of existing malaria control programs and the construction, equipping, and staffing of municipal health centers.[16] The second World Bank project, the São Paulo Basic Health Project, provided US$55.5 million to improve cost-effectiveness of health services in greater São Paulo with the assumption that this measure would improve residents' health. The project sought to help decentralize the provision of health services away from the state by creating localized, yet integrated "health modules" that offered a variety of places where people could seek medical services. According to the World Bank, such modules would encourage people to use clinics by offering lower prices for basic care than a hospital would charge. Both projects resulted in the construction or upgrading of a number of hospitals and basic care clinics, but the World Bank reported that many of the clinics were understaffed or only occasionally open, or both. As a result, those seeking medical aid often still went to the hospital, and the goal of achieving more cost-efficient care was not achieved.[17]

Despite such failures, ratification of the new Brazilian constitution in 1988 helped institutionalize the idea that decentralized health care would provide more efficient and effective health services.

In 1988, health care was established as a universal right and guaranteeing that right became the responsibility of the state. Prior to the passage of the new constitution, the state health care system for urban private sector employees was funded through their contribution to the social security system. However, since many Brazilians worked in the informal sector or held temporary jobs, the system was far from universal. To rectify this problem, the new constitution called for the creation of the Unified Health System (Sistema Único de Saúde, or SUS) to oversee "a regionalized and decentralized network of health services."[18] Decentralization entailed municipalizing health care in Brazil. In this context, the Unified Health System moved beyond constructing and upgrading medical facilities to, most notably, expanding health care through the Family Health Program. Introduced in 1994, the program sought to create core teams of health professionals to provide services to a defined geographic area consisting of up to a thousand households. The teams usually were made up of a physician, nurse, nurse assistant, and four to six community health workers. These teams provided health care to individuals at their local clinic. Community health workers also made monthly visits to their constituents' homes to ensure they were able to access care when needed.[19]

Decentralization affected not only how Brazil administered health care but also how the country financed it. In the 1980s, the federal government, through social security, funded more than 75% of health care spending. In compliance with IMF and World Bank structural adjustment policies in the mid-1990s, however, the federal government was pressed to decrease its financial contribution to health care, and its share dropped to close to 50%. To make up the difference, financial responsibility for health care was placed on states (12%) and municipalities (15%).[20] In this context, the World Bank could more easily circumvent the federal state. From 2000 to 2015, only 28% of World Bank–funded projects focused on the federal state, while 72% focused on subnational entities. In addition, the World Bank favored funding projects in places where Unified Health System managers had greater decision-making power and autonomy, because it could more easily influence health care policy there in ways that aligned with its ideological focus on economic efficiency and the private provision of health care services.[21]

The decentralization and reorganization of health services in Brazil were accompanied by the decentralization and reorganization of vector-borne disease control efforts. In 1990, Brazil moved the responsibilities of the National Epidemiological Surveillance System (Sistema Nacional de Vigilância Epidemiológica, or SNVE) to the newly created National Health Foundation (Fundação Nacional de Saúde, or Funasa).[22] Broadly speaking, Funasa was originally created to aid with the implementation and expansion of the Unified Health System. Upon absorbing the responsibilities of SNVE, Funasa was also responsible for combating dengue. Throughout the second half of the 1990s, Funasa helped implement the Ministry of Health's *Ae. aegypti* eradication plan. Funasa later shifted its focus from eradicating to controlling the mosquito through an intensified plan to control dengue in 2001 and creation of the National Program for Dengue Control (PNCD) in 2002. The control plans and programs reflected a more holistic approach to health in that they sought to integrate epidemiological surveillance, vector control, sanitation, and primary health care.[23]

As with the provision of health care, responsibility for implementing Funasa's dengue programs was decentralized and delegated to municipalities. For the eradication plan, this entailed persuading local municipalities to agree to implement Funasa's recommendations. While some municipalities signed agreements soon after the plan's initial release, others took several years, and still others never did. More than a billion Brazilian reais were allocated among the municipalities that did sign on. Most of the funds were used for hiring and training workers and purchasing equipment. More expensive needs, such as sanitation improvements, often went unmet. For the dengue control plans, the funds were more targeted to municipalities that had higher risks of dengue outbreaks. The federal government's funding allocation resembled that of the eradication plan, but moneys went largely to disease and vector surveillance and control.[24]

Creation of the Unified Health System had a positive impact on health in Brazil as a whole, but the decentralization of government services and responsibilities more broadly continued to localize disease in individual bodies and proved ineffective at controlling the *Ae. aegypti* and the diseases it carries. Constitutionally mandated to provide health care as a universal right, the Unified Health System and municipalization of health services did widen access to health care, particularly in impoverished and previously underserved communities.[25] In addition, Brazil's dengue control plans recognized that outbreaks of the disease were linked to broader social and environmental factors. However, Funasa's efforts to abate the presence of the *Ae. aegypti* had limited success, and the Unified Health System proved only partly able to treat or stave off mosquito-borne diseases. Due to budgetary constraints, limited institutional capacity, and political priorities, municipalities often focused on control measures that had more immediate and shorter-term impacts. Municipal actions entailed medically treating individuals with dengue and targeting mosquitoes with larvicides and pesticides, rather than expanding or improving water, sewage, and sanitation infrastructure. Paid for with public funds, these measures focused largely on changing individualized and small-scale environments—the human body and the home. As a result, the number of dengue cases and dengue-caused deaths steadily rose.[26]

Decentralization thus may have widened access to health care, but it resulted in disjointed and inadequate responses to controlling mosquitoes and the diseases they carry. In the words of Nilson do Rosário Costa, a researcher with Brazil's principal public health institution, *Ae. aegypti* "has no owner," yet Brazil's dengue control programs have made it the responsibility of municipalities.[27] Through decentralization, each municipality has become responsible for an array of activities, including collecting and analyzing the serological samples of patients, treating patients with severe cases of dengue, tracking disease occurrences and outbreaks, and conducting entomological surveillance, mosquito control, and the oversight and expansion of sanitation services. However, the ways each municipality has dealt with such responsibilities has varied greatly. "In metropolitan areas," do Rosário Costa observes, "the success of a municipality in controlling dengue does not necessarily produce sanitary safety because nothing guarantees that the neighboring city has also actually committed to the problem."[28] In the long term, investments in infrastructure for water, sewage, and sanitation would likely have greater, longer-lasting impacts on mosquito-borne diseases.

## THE CREATION OF VALUE IN A MOSQUITO

While the *Ae. aegypti* technically "has no owner," Oxitec sought to make the mosquito into something that could be bought and sold. Three British scientists founded Oxitec in 2002 as a "spin-out" (or, in US English, a spinoff) from a lab at the University of Oxford. Lead scientist Luke Alphey and his colleague David Kelly worked in Oxford's Department of Zoology. The third scientist, Paul G. Coleman, was affiliated with the London School of Hygiene and Tropical Medicine. Alphey had worked on insect genetics for decades. In the 1990s, Alphey and his team became interested in developing a gene-driven autocidal insect technique, what eventually became known as "release of insects carrying a dominant lethal" (RIDL).[29]

The idea for RIDL emerged from the sterile insect technique (SIT), first developed in the early twentieth century. Scientists found that exposing insects to gamma and X-rays resulted in sterility or dominant lethal mutations that debilitated the insects' offspring. To control insect populations, male insects subjected to SIT were released into the wild. When they mated with wild female insects, their offspring frequently died before or shortly after maturation. By the 1950s, this technology was being successfully used in integrated pest management programs to control for the screwworm fly, a deadly livestock parasite. SIT has also been used in different parts of the world to eradicate both the melon fly and the Mediterranean fruit fly. In addition, SIT fruit flies have been used preventively at different ports in the United States to control pests that arrive in or on imported goods. The use of the technology to control for mosquitoes was first experimented with in the 1960s. In the 1970s, the technique was shown to be somewhat effective in the control of *Ae. aegypti* populations. However, the SIT mosquitoes occasionally proved, on release, not to compete well with their wild counterparts in mating.[30]

While SIT and RIDL are used to effect similar results, the process behind the two techniques differs. Both seek to decrease insect populations through dominant lethal mutations and thus decrease the likelihood of offspring survival. However, whereas SIT uses radiation to mutate an insect's genes, RIDL uses genetic engineering. RIDL insects are made to have mutations that are repressible through a nonnaturally occurring chemical additive. Alphey and his team developed RIDL insects whose dominant lethal mutation was repressed when tetracycline was added to their diet. In addition, Alphey's RIDL insects contained a fluorescent marker that was passed on to their offspring so researchers could determine whether the technique worked. To make the technology usable outside the lab, male RIDL mosquitoes were fed a diet that included tetracycline and then allowed to breed with female non-RIDL insects in a controlled environment. With the autocidal gene repressed, the male insects survived but still passed on their mutated genetic code to the eggs of their progeny. The next generation was then reared to pupae, when they could then be sorted by sex. As the sex that bites humans and carries dengue, the female pupae were terminated. The male pupae were removed from the controlled environment, introduced to a place conducive for them to hatch, and then freed to breed with wild female *Ae. aegypti*. The insects born from this second-generation progeny were debilitated and not able to live to a reproductive age.[31]

Isis Innovation, Oxford University's technology transfer and consultancy arm, first saw the value of the RIDL technique in early 2000s. Working with Alphey and his team, Isis Innovation provided the seed funding for Oxitec as a university spin-out. Spin-outs emerged in the United Kingdom as part of the broader push to privatize and financialize higher education and scientific research in the face of decreasing public funding. In contrast to more traditional forms of managing intellectual property, spin-outs were developed to find investors for early-stage scientific and technological discoveries that were still in development and not yet patented. For universities, spin-outs were seen as a way to decrease costs. Involving investors earlier in the research process redirected the costs of research funding and patents away from universities. Spin-outs like Oxitec also increased the profits universities and scientists could potentially make from their discoveries by making both into vested shareholders of a company that could later be sold. At Oxford, the university usually takes 50% of the initial shares in spin-outs, and the scientists take the other 50%. Once investors are found, more shares in the spin-out are added. The new shares are then divided between investors at an agreed-on proportion.[32]

By the time Oxitec was started in 2002, Alphey and his team had demonstrated success in the late 1990s with RIDL technology in fruit flies.[33]

They turned to mosquitoes, and by 2005, preliminary lab results showed that the RIDL system worked in *Ae. aegypti*. Four years later, the first field test of the mosquito—known as OX513A—demonstrated that male RIDL mosquitoes were as competitive as their wild counterparts in mating and passing on the RIDL autocidal gene.[34] Further studies demonstrated that sustained release of the mosquito suppressed local populations of *Ae. aegypti*.[35] The RIDL technique was later expanded to target a number of other insects. In 2016, Oxitec filed to trademark RIDL under the Friendly™ moniker.[36]

In many ways, Oxitec's Friendly™ technology was an ideal way to turn a mosquito into a commodity. First, the OX513A mosquito was on the surface easily marketable. The expanding range of the *Ae. aegypti* and the increasing threat of vector-borne diseases such as dengue and Zika had brought greater attention to mosquito control. Unlike pesticides, OX513A offered a solution that reportedly posed no harm to humans yet still limited the presence of the *Ae. aegypti*.[37] Second, OX513A was less expensive to make than sterile insects. Compared to traditional SIT, the OX513A had lower overhead and was more likely to be effective. Whereas SIT required using radiation and then releasing live mosquitoes, the OX513A involved only breeding mosquitoes for their eggs. This method lowered both input and transportation costs of the insect while it also increased the mosquitoes' survival and successful mating rates. In addition, since *Ae. aegypti* eggs can remain viable for months, OX513A had a much longer shelf life than the few days of traditional SIT mosquitoes. Third, the use of OX513A to suppress local populations also depended on its continual release. Its initial use required a number of OX513A large enough to be released for a long enough period of time to outcompete and outlast wild populations.[38] While field studies have demonstrated that the use of OX513A significantly decreased local *Ae. aegypti* populations, the wild mosquito population has been shown to rebound within only a few months after discontinuing release of the mosquitoes.[39]

The commodifiable traits of the OX513A and other Friendly™ insects made Oxitec into a highly valued company. In 2015, the sale of Oxitec to Intrexon proved quite lucrative for those who held shares in the company. In 2015, Oxford University Innovation reported that it had invested around US$384,000 in seed funding in Oxitec.[40] According to Vice-

Chancellor Dr. Andrew Hamilton, Oxitec netted the university a return on investment of US$14.4 million. Oxford's return derived in part from its initial shareholdings in the spin-out and in part from having increased its stake in Oxitec through Oxford Spin-Out Equity Management (OSEM).[41] According to US Security and Exchange Commission filings, the scientists and cofounders of the project also netted a sizable return. Alphey made US$9.2 million off the sale of his shares, receiving half in cash and half in Intrexon stock. In addition, he received US$1.3 million from the sale of his options, split equally between cash and stock. Kelly made US$1 million and Coleman made US$38,119 from their respective shares, with the earnings also split equally between cash and stock.[42]

OX513A and Oxitec's Friendly™ technology thus appeared to be profitable commodities, and the company's original shareholders gleaned a sizable return from the company's sale. But the company had funded most of its operations with government agency grants and investments by venture capitalists and charitable foundations. The Wellcome Trust operated in both later roles, investing close to US$1.3 million in the company before 2015. When Oxitec was sold to Intrexon, the Wellcome Trust received US$1.68 million in cash and a matching amount in stock. In 2005, Oxitec also received US$19.7 million from the Foundation for the National Institutes of Health in the United States and a US$4.8 million grant from the Grand Global Challenges Initiative, led by the Bill & Melinda Gates Foundation.[43] Wellcome Trust and the Bill & Melinda Gates Foundation both continued to support Oxitec after its sale. Wellcome Trust provided the company with a US$6.8 million dollar grant in 2021. The Bill & Melinda Gates Foundation provided the company with a US$4 million grant in 2018, a US$1.6 million grant in 2020, and a US$18 million grant in 2022.[44]

This steady stream of funding kept Oxitec afloat, but the long-term value of OX513A depended on the sale and ownership of an idea. That idea was bought and sold as shares in Oxitec and materially encapsulated in the company's Friendly™ technology. While the Friendly™ line of products may seem promising, the sale of the insects as profitable commodities was not realized in the first two decades of Oxitec's existence. However, the continuing threat of dengue, Zika, and other viruses carried by the *Ae. aegypti* in Brazil, combined with the country's public yet decentralized health care system, provided Oxitec with a place to test both the efficacy and the potential profitability of the Friendly™ *Ae. aegypti*.

## THE REALIZATION OF VALUE IN A MOSQUITO

For Oxitec, Brazil offered an ideal place in which to realize the value of the Friendly™ *Ae. aegypti*. The country had a long history of using scientific advancements and working with outside organizations to combat *Ae. aegypti* and other insect populations. In addition, dengue rates had steadily risen for decades. Brazil was thus a place where there was a seeming need for the Friendly™ *Ae. aegypti*, along with a scientific environment and some existing infrastructure to make its use possible. But perhaps most important, Brazil's decentralized public health and insect control system could provide Oxitec with multiple potential openings and avenues for making a profit from its genetically modified mosquito, or at least appearing to do so, for the company's shareholders.

While OX513A was deemed a laboratory success in 2005, global regulation of the transport of living modified organisms (LMOs) slowed Oxitec's ability to release and test the mosquito's effectiveness in field trials. The Cartagena Protocol on Biosafety limits the export of LMOs that are intended to be released in noncontrolled environments without a prior risk assessment. While Oxitec conducted such an assessment for the OX513A mosquito, the company initially exported the mosquito's eggs to receiving countries for contained use. This enabled Oxitec to rear the Friendly™ *Ae. aegypti* in local labs and develop locally produced egg supplies. This step relieved the company from having to provide importing countries with risk assessments. Once the OX513A were hatched from eggs laid in labs in importing countries, the mosquito's release was subject only to national regulations. Oxitec performed its first field trial in the Cayman Islands. As a British protectorate, the Cayman Islands is not a member state of the United Nations and not required to abide by UN protocols. To release OX513A in field trials, Oxitec thus worked with the Mosquito Research Control Unit in the Cayman Islands to obtain the necessary national permits. The company then conducted its first field trial in the Caymans in November 2009. In December 2010, Oxitec performed a second field trial in Malaysia. The company followed its earlier procedure and first imported the eggs to the country for contained use. However, as a UN member, Malaysia abided more closely to the Cartagena Protocol and, before the field trial was to begin, formally notified the Biosafety Clearinghouse of the mosquito's planned release into an uncontrolled environment.[45]

In 2009, Oxitec also sought to bring OX513A to Brazil. To help facilitate the process, Oxitec recruited local partners that had prior experience with sterile insect techniques (SIT) and *Ae. aegypti* eradication to work on what became known as the Projeto *Aedes* Transgênico (PAT). The Projeto *Aedes* Transgênico was launched by Moscamed, a nonprofit organization largely funded by the Brazilian Ministry of Agriculture.[46] Moscamed had used traditional SITs to control Mediterranean fruit flies on local agriculture.[47] In addition to Oxitec, the PAT included researchers from the University of São Paulo. The Oswaldo Cruz Foundation was also initially asked to join the partnership but declined after Oxitec refused to transfer its RIDL technology to local Brazilian partners. The Projeto *Aedes* Transgênico was also supported by the UK's Department of Trade and Industry, which seeks to enhance exports from British companies.[48]

Oxitec and its partners deployed OX513A in Brazil much as it had elsewhere, importing the mosquitoes' eggs to first be reared in labs and then later be released into the environment. To bring the mosquitoes' eggs to Brazil, Oxitec needed to receive approval from the Brazil's National Technical Commission for Biosafety (Comissão Técnica Nacional de Biossegurança, or CTNBio), a governmental advisory body that reviewed applications for the introduction of genetically modified organisms (GMOs). Previous to Oxitec's application, CTNBio had approved only the use of GMO crops. In September 2009 the agency approved Oxitec's request to import three batches of 5,000 OX513A eggs each from the company's lab in the UK for a contained release in Moscamed's existing facilities in Juazeiro, Bahia. After OX513A was reproduced at the Moscamed facilities, Oxitec filed a second request with CTNBio, in December 2010, to deploy the mosquito into uncontained environments. Three months later, the first OX513A mosquitoes were released in the suburbs of Juazeiro. By the time Oxitec's eighteen-month field trial concluded, over 15 million OX513A had been released into the environment.[49] According to the principal scientists working on the project, the release of OX513A resulted in a 95% reduction in the number of adult *Ae. aegypti* in the study area.[50] After the success of this field trial in Juaziero, Oxitec expanded its operations southward in Bahia to Jacobina. There it opened a new OX513A-rearing facility with the intention of expanding the field trials to a more densely populated urban setting. In June 2013, Oxitec commenced a plan to release 4 million OX513A a week for two years in the city.[51] While the numbers reported after the field trial put weekly releases closer to 450,000 mosquitos, scientists reported a level of efficacy similar to those of previous studies in reducing the *Ae. aegypti* population.[52]

The field trials in Bahia were intended not only to demonstrate the scientific efficacy of Oxitec's Friendly™ *Ae. aegypti* but also to bolster the insect's profile as a safe and, in turn, profitable commodity. During the trials, the mosquito drew praise in the international press and from politicians. The BBC World Service called the OX513A "a revolutionary new approach" that would deliver "a killer blow to the dengue-carrying mosquito."[53] In a visit to Juazeiro in April 2011, the governor of Bahia also lavished Oxitec with support, and the state secretary of health personally released close to a thousand OX513A. The project in Jacobina received a similar level of attention. An array of federal and state level officials were present at the opening of Moscamed's new facility, including the Brazilian minister of health. The minister of the UK Department of Trade and Investment also formally congratulated Oxitec "for securing new partnerships in Brazil" and for being "a crucial part of the Government's plan for growth."[54]

After the completion of the field trials in Bahia, CTNBio approved the OX513A as safe for commercial release in April 2014. As an advisory body charged with evaluating the potential risks only of GMOs, CTNBio focused only on whether the OX513A would have adverse effects on humans and the environment. The agency explicitly noted that its approval "does not focus on the technological efficacy, costs, and advantages or disadvantages of other technologies of *Ae. aegypti* population control." Furthermore, the agency made clear, "questions directly linked to dengue control do not concern CTNBio" but instead rested with "the [Federal] Ministry of Health and [State] Secretariats of Health." In this respect, CTNBio left it to the Brazilian Health Regulatory Agency (Agência

Nacional de Vigilância Sanitária, or Anvisa) to comment on the OX513A's effectiveness as a public health intervention and, in turn, to provide the commercial registration necessary for the mosquito to be bought and sold. Until Anvisa approved the use of OX513A, the mosquito could be used only in additional scientific field trails. In April 2016, Anvisa approved OX513A for commercial use but on a temporary basis as long as those deploying it continued to track its presence and effectiveness.[55]

To enhance the OX513A's future use and profitability, Oxitec used the publicly funded, yet decentralized Brazilian health care system to expand the legitimacy and use of its Friendly™ mosquitoes. In July 2014, Oxitec opened its own OX513A production facility in Campinas, São Paulo. The facility not only allowed Oxitec to operate independent of its Brazilian collaborators but also provided the company with an inroad into the country's wealthiest and most populated region. Later that year, Oxitec worked out an arrangement to deploy its Friendly™ *Ae. aegypti* in the nearby city of Piracicaba. In particular, the company sought to release the mosquitoes in the city's Eldorado neighborhood, where the incidence of dengue was high. While local concerns about the health risks of GMO mosquitoes temporarily delayed the trial, Oxitec worked with community health workers to quell residents' fears. As an existing part of Piracicaba's public health care structure, community health workers were already making regular visits to people's homes. To legitimate the mosquito's presence, Oxitec sent representatives of its team along on the home visits to promote OX513A's safety and effectiveness.[56]

Oxitec also tapped into Piracicaba's public health and dengue eradication funds. After Oxitec's initial trial, the Piracicaba Epidemiological Surveillance Service reported a significant drop in dengue cases in the Eldorado neighborhood. The decentralization of Brazil's public health care and mosquito control programs devolved on municipalities a degree of discretion over how the funds for such programs are used, so the city agreed to help pay for an expanded release of OX513A to eleven downtown districts populated by about 60,000 people. While the mosquitoes technically could not be sold, the city of Piracicaba agreed to make what Oxitec called an US$800,000 "contribution" to an "optimization study."[57] Neither the city of Piracicaba nor Oxitec reported the total cost of this "study." However, officials in Piracicaba publicly stated it would cost about

R$30, or approximately US$7.50, a year per person protected. With a population of 390,000, the bill for the city was about US$2.7 million a year for the release of close to 3 billion OX513A annually. The estimated price of the Friendly™ mosquitoes would about equal the cost of the pesticides, larvicides, and sick leave the Piracicaba Health Department already paid to control and treat dengue.[58]

In 2018, Oxitec began to take additional measures to fully register its Friendly™ mosquitoes for commercial use. After years of delay, the company filed a lawsuit against Anvisa to persuade the agency to approve the application permanently.[59] Oxitec argued that CTNBio had already acknowledged that the mosquitoes posed minimal to no risks to human or environmental health when compared to the same nongenetically modified species. In March 2018, the judge overseeing the lawsuit suspended Anvisa's review and stipulated that Oxitec could commercially sell its Friendly™ mosquitoes. As an agency designed to approve the health-related claims of companies seeking to sell products in Brazil, Anvisa responded that it had not approved Oxitec's request because evidence was insufficient that the Friendly™ *Ae. aegypti* could effectively control dengue, Zika, and chikungunya. The judge based the court's decision to overrule Anvisa on "research" obtained on the "world wide web" which mentioned the success of Oxitec's mosquitoes in the Piracicaba.[60] At the time of the decision, no peer-reviewed research had been published on either the health benefits of the Friendly™ mosquitoes or on Oxitec's study in Piracicaba. The documents referred to by the judge were thus either Oxitec's press releases or articles quoting Oxitec's press releases or employees.

In May 2018, Oxitec announced its first field trial for a second generation of Friendly™ *Ae. aegypti* in Indaiatuba, São Paulo. The new mosquito—the OX5034—was genetically engineered to be male selecting, meaning that the autocidal gene kills only female offspring. The male offspring live and still carry a copy of the male-selecting autocidal gene. In addition, their offspring continue to pass on the autocidal gene to half of their own progeny. The OX5034 thus potentially limits the number of mosquitoes needed to decrease *Ae. aegypti*'s populations by having a multigenerational (tested up to ten generations) yet still self-limiting suppression effect. Since the females die, the OX5034 also does not need to be sex sorted before being released outside the laboratory. In addition, its eggs can be brought to and hatched in the places where the mosquito is going to be used.[61] Oxitec's OX5034 trials, the company reported, demonstrated a 96% peak suppression rate, nearly eliminating local populations of *Ae. aegypti*. In its press release, the company's CEO compared the technological leap from the OX513A to the OX5034 to that "from the original Model T car . . . to a self-driving electric vehicle."[62]

In 2020, CTNBio approved the second-generation Friendly™ *Ae. aegypti* mosquito for commercial use. While Oxitec continued to work with municipalities to incorporate the OX5034 into their insect control plans, it also pursued another avenue to help realize the value of the mosquito: the "Friendly™ box." To control for *Ae. aegypti*, an egg pack just needs to be mixed with water in the box and then replenished on a monthly basis.[63] In November 2021, the Friendly™ box became available for online purchase in Brazil on a subscription basis: for R$145 (approximately US$25) a month (Visa, Mastercard, or American Express accepted), a Friendly™ box is sent directly to the customer's doorstep.[64]

## FRIENDLY™ SHAREHOLDER VALUE

In Oxitec's initial press release about the OX5034 and the Friendly™ box, the company's CEO, Grey Frandsen, spoke proudly of the product, promoting it as a way for communities and individuals to safely take action to control dengue. "We've placed the power of Friendly™ biology into a small, joyful box," Frandsen said, "giving everyone . . . the ability to control the dengue-spreading *Aedes aegypti* without chemical pesticides. . . . It's time to give power to the people to act against this growing public health threat with safe, effective products they will love to use."[65] Frandsen's comments made the Friendly™ box seem a valuable product and a win for individuals, communities, and the environment. However, examining how the Friendly™ box and Oxitec's mosquitoes are embedded in the broader history of global finance tells a different story.

The groundwork for turning the Friendly™ *Ae. aegypti* into a commodity and selling it in Brazil was laid in the 1980s and 1990s. As governments in the global north cut financial support for higher education, the purpose of scientific discovery changed from the pursuit of knowledge in the interest of the public good to the pursuit of profit. At the same time, global financial institutions such as the World Bank and IMF used structural adjustment programs to promote the decentralization of state administrative systems in order to create new investment opportunities for companies from the global north under the guise of economic efficiency. Taking this into account, scientists from universities such as Oxford sought to create salable commodities like Friendly™ insects, as well as salable companies like Oxitec. To find a market for their commodities, such companies took advantage of newly decentralized administrative systems. In places like Brazil, Oxitec could negotiate more directly with municipal governments to use the Friendly™ *Ae. aegypti* as part of their insect control plans. In addition, the company was able eventually to develop a direct-to-consumer model for its genetically modified mosquito and thus potentially increase product demand.

The commercial release of the Friendly™ box may have given people the power to purchase the product, but the joy and safety it may bring is likely limited. At BRL145 a month, the Friendly™ box is out of reach for those in Brazil most impacted by the burdens of *Ae. aegypti*-borne disease, particularly the poor in the underserved areas where the mosquito was tested. The Friendly™ box would cost a household earning in the bottom quintile of Brazil's population somewhere between 20% and 25% of its monthly income. Depending on Friendly™ mosquitoes as the sole means of controlling yellow fever, dengue, and Zika could also be ecologically fraught. The release of the genetically modified mosquito has been demonstrated to decrease the population of *Ae. aegypti*, but other types of mosquitoes in Brazil can carry the pathogens that cause dengue and Zika. Most notably, the *Aedes albopictus* serves as a vector for both diseases, albeit with lower rates of infection and transmission than the *Ae. aegypti*.[66] In addition, different types of mosquitoes have been shown to lay their eggs in similar places where their larvae then compete for nutrients to survive. In the process, the mosquitoes limit one another's ability to reach adulthood in full health.[67] If the presence of one type of mosquito decreases, another may increase or advance to adulthood with more vigor. Using the Friendly™ *Aedes* instead of investing in infrastructure to reduce standing water and instead of using pesticides or larvicides could plausibly result in an increase in the presence of other types of mosquitoes and thus blunt the impact of the genetically modified mosquitoes' ability to help control disease. As Marcelo Jacobs-Lorena, a molecular entomologist at Johns Hopkins University, notes, the use of the Friendly™ *Ae. aegypti* to suppress mosquito populations "may be good for the company, but maybe not so good for overall control efforts."[68]

While Oxitec and its Friendly™ *Aedes* could potentially live up to the company CEO's lofty statements about the product, their value rests more in turning a profit than in sustaining public health. After Intrexon purchased Oxitec in 2015, the price of Intrexon stock steadily decreased. By late 2018, the company's share price dropped below US$7.[69] Around that time, Intrexon decided to rebalance its portfolio to focus more squarely on pharmaceutical and medical advancements. In the process, Randal J. Kirk stepped aside as CEO of the company but was soon appointed executive chair. Along with this change, Kirk's private investment firm Third Security agreed to purchase Oxitec and Intrexon's non-health-care-related assets for US$53 million in cash and with an agreement to purchase US$35 million in Intrexon common stock.[70] The total amount spent by Third Security to obtain the suite of assets was only slightly more than half of what Intrexon originally paid to acquire Oxitec alone. Third Security thus acquired Oxitec and its Friendly™ technology as heavily discounted assets.[71] To see a return on the investment, both the firm and its insects need to be made to look like products that can be bought and sold in the global marketplace.

# 6 Treating from Home

On May 23, 1967, Chairman Mao Zedong, the leader of the People's Republic of China, launched the top-secret Project 523. The code name reflected the date the project was initiated. Chairman Mao dedicated the labor of five hundred Chinese scientists to Project 523, whose aim was to find a new cure for malaria.[1] One group of Chinese scientists started their work by combing through texts on traditional medicines and other ancient Chinese medical writings. They soon stumbled upon a fourth-century text noting that *qinghao*, or what the West calls *Artemisia annua*, could be used as a treatment to alleviate malarial fevers. In more common vernacular, *qinghao* or *Artemisia annua* is known as sweet wormwood, a spiky-leafed plant with yellow flowers. Further research by the Chinese scientists found that the plant and its medicinal properties was etched on tomb carvings as far back as 168 BCE and mentioned in many medical scrolls thereafter.[2]

While the ability to use *qinghao* to treat malaria appeared in an English-language scientific publication in 1979, it was not until China filed a patent on artemisinin drugs in 1990 that it drew the interest of Western pharmaceutical companies.[3] In 1991, a partnership was formed between Chinese researchers and the Swiss drug manufacturer Ciba

Geigy, now named Novartis. Three years later, Novartis patented a pill containing an artemisinin derivative and launched its artemisinin combination therapy (ACT) treatment in 1999. At the time, a single dose of the most common treatment for malaria, chloroquine, cost less than 25 cents. Novartis priced its new drug at US$44 per course. With the effectiveness of chloroquine waning as a malaria preventive and treatment, Novartis's ACT treatment was marketed to wealthy European and American travelers going on safari in Africa. Novartis eventually released a cheaper version of its ACT, named Coartem, to be sold and used in lower-income countries. However, priced at US$2 per pill in 2021, it was still several times the price of chloroquine and difficult for many of the people most impacted by malaria to afford.[4]

By the time Kelly first visited the Bududa District in Uganda, in 2013, Coartem had become a mainstream malaria treatment. Generic versions produced mainly in India were also easily available in Uganda at "drug shops," the term many in Uganda use to refer to the local pharmacy. At different points in time, the Uganda government had subsidized the price of Coartem and occasionally provided it to people for free. However, malaria was and still is pervasive in the Bududa District. A day rarely passes without confronting the disease in some way. The members of the local family that Kelly stayed with often suffer from bouts of malaria while she is there. In almost every interview or interaction with community members, malaria finds its way into the conversation. Many people talk about how many times they have fallen ill with malaria in the past year or about the children and other family members they have lost to the disease. While first-line malaria drugs can now be bought at any of the local pharmacies, a common consensus has it that the malaria problem in the community is getting worse. Community members often note that over the past several years "malaria has increased a lot." Some talk about "going to the health center and seeing how many people are lining up for treatment. It is many, so many." Others comment on the lack of effectiveness of malaria medications, saying they think rates of malaria have gone up because "the medication is not so tough to treat the disease."

The people in Bududa are thus getting hit from both sides. Over the past several decades, the World Bank and IMF have conditioned their loans to Uganda on privatizing and liberalizing the country's health care system; in the process, the country's public health infrastructure has degraded. In addition, with the most effective malaria treatment originally patented and marketed to boost shareholder value, and with generic versions today still costing several times more than other forms of treatment, the best malaria medicines are still out of reach for many of the people most in need. Financialization has thus not only led to the use of coffee-farming practices that are cultivating mosquito habitats, but also constrained the availability and accessibility of the most effective forms of malaria treatment.

## THE PATHOGENS OF MALARIA

In 2020, there were 241 million malaria cases and 627,000 malaria deaths worldwide. While nearly half the world's population lives in malaria-endemic zones, more than 95% of malaria cases and deaths are concentrated in sub-Saharan Africa. Five countries—one of which is Uganda—account for more than half of all malaria cases and deaths worldwide. The number of global malaria cases has been slowly rising since around 2016. The World Health Organization attributes some of the increase in the most recent years to disruptions in malaria prevention and treatments due to COVID-19. The WHO also changed its methodology for estimating malaria cases in thirty-two countries in sub-Saharan Africa. The methodological changes revealed that malaria has taken a considerably higher toll on people in the region than previously thought. Further, the WHO attributes the disease resurgence that sub-Saharan Africa is experiencing to more than changes in counting.[5]

Like many vector-borne diseases, malaria takes time to manifest in the human body. After being bitten by an infected *Anopheles* mosquito, a person inflicted by *Plasmodium falciparum* parasites may not display symptoms until two weeks later. Most people suffering from malaria experience a fever, headache, muscle aches, fatigue, and stomach discomfort. In more severe cases, malarial infections can cause kidney failure, seizures, mental confusion, and coma. Malaria can quickly become fatal if not promptly treated, especially in young children, pregnant women, and others who have not regularly and recently been infected. Once symptoms appear, a mosquito that feeds off an infected human host can pass the parasite on to its next victim. The WHO thus recommends that all suspected cases of malaria be confirmed using parasite-based diagnostic testing, through either microscopy or a relatively newer innovation, rapid diagnostic tests.[6] Early diagnosis and treatment of malaria is crucial, not only to protecting a person's health but also to preventing malaria from spreading to others.

However, in places like Bududa, most people never get a formal diagnosis and instead "treat from home." An underfunded health care sector has left the clinics used by those most susceptible to malaria without the necessary medical staff, equipment, or stable sources of electricity to adequately test people for malaria. In addition, such clinics lack steady supplies of malaria medications that can be given to those who test positive for the disease. Households in Uganda are thus often forced to seek frontline malaria therapies at their local drug shop and then self-treat. Without affordable access to proper medical advice or the most effective malaria medications, many people cut corners in their own treatment to save money. However, taking an inadequate dosage of a malaria medication or taking it for a shorter duration makes the disease more drug resistant and, in turn, more difficult to eradicate.

## MALARIA MEDICATIONS AND THE MOTIVES OF FINANCIAL CAPITAL

Malaria has impacted people's health and likely been a contributor to premature death since the dawn of human civilization. The quest to prevent and treat malaria is thus centuries old. One of the earliest known treatments for malaria came from the bark of cinchona trees. Cinchona trees were originally found in the high-cloud forests on the eastern slopes of the Andean Mountains, where Indigenous peoples are thought to have long used their bark to treat fevers and diarrhea. In the 1600s, Jesuit missionaries brought the bark to Europe, where it was found to be a useful treatment for malaria. Shortly thereafter, the cinchona tree became an object of desire among European colonizers engaging in wars and seeking to expand their imperial footprints in malaria-ridden parts of the world. With no known alternative treatment for malaria, European countries thus began cultivating the cinchona tree throughout their colonies.[7] In the early eighteenth century, scientists isolated the active ingredient in cinchona bark that is now known to cure people of malaria: quinine. With no way to artificially produce the substance, however, the cinchona tree remained the only known source of quinine in the world.

It was not until 1934 that German researchers working for Bayer developed an alternative synthetic malaria treatment: chloroquine. The drug remained in trial phases through the beginning of World War II. In 1943, Allied forces learned of the drug and sent it back to the United States for further investigation and development.[8] Soon thereafter, chloroquine-based medications became the dominant treatment for malaria. Compared to quinine, the drug was more effective, its effects lasted longer, and it was relatively easy and cheap to produce.[9] For these reasons, physicians and WHO experts considered it a "presumptive treatment" not only for malaria but for a wide array of other ailments, recommending that people take it at the first sign of a fever.

The effectiveness of both quinine and chloroquine as malaria treatments began to wane during World War II. The *Plasmodium* parasite that causes malaria is remarkably resilient and adaptable. Every species of the malaria parasite has numerous genetically distinct strains, each with a unique set of strengths and weaknesses. As both quinine and chloroquine became increasingly available without regulated or proper use, resistance to the two treatments emerged. As the widespread use of pesticides and antimalarial treatments resulted in the eradication of malaria in the global north in the 1950s, such resistance was not of immediate concern in many industrialized countries in the world. However, during the Vietnam War, it became apparent that both Vietnamese and American fighters were dying of malaria despite the use of chloroquine.[10] In response, Vietnam appealed to China for help. Although Chairman Mao was at the time overseeing the Cultural Revolution, in which scientists and intellectuals were persecuted by the Communist Party, aiding the Vietnamese soldiers was in Mao's political interests. In addition, many people in southern China who lived in subtropical zones were also suffering from malaria. Vietnam's request resulted in the commencement of Project 523 and the eventual rediscovery of artemisinin as a viable malaria treatment.

In 1972, the Chinese scientists working on Project 523 identified artemisinin as the key active component of the *qinghao* extract.[11] In testing the artemisinin derived from the plant, they found it has a unique structure. In technical terms, it is a sesquiterpene lactone with a peroxide bridge. In lay terms, this means that artemisinin killed malaria parasites faster, with less toxicity, and in a completely different way than quinine or chloroquine. This unique chemical structure meant that artemisinin could be used in places where *Plasmodium* parasites had developed resistance to other malaria treatments. Early trials that Chinese scientists conducted using mice showed that artemisinin was 100% effective when used to treat malaria.[12] Human trials later showed similar results, but they also showed that the body rapidly eliminates artemisinin. Any parasites that survived the treatment thus made a quick comeback. Later studies demonstrated that artemisinin derivatives used by themselves as a monotherapy had a relatively high relapse rate of about 10%, and that *Plasmodium* parasites could quickly develop resistance to the new treatment. As a result, scientists mixed artemisinin with slower but more persistent drugs to develop what are called artemisinin combination therapies (ACTs).[13]

The discovery of artemisinin was initially considered a military secret, but even after it was declassified, knowledge of it received little attention. Its use as a malaria treatment was reported in scientific publications in Chinese in 1977 and in English in 1979. From 1980 to 1990 the use of artemisinin drugs in China reduced the annual number of malaria cases from 2 million to 90,000.[14] Western scientists doubted the validity of the research underpinning the drug's use and success; in particular, they claimed that the research and associated trials were conducted with rudimentary equipment not up to World Health Organization standards. Distrust between China and the West and China's isolationism also played a role. In addition, patent laws were nonexistent in Communist China. This meant that anyone could use artemisinin and, in turn, there was no way for a major drug company to get a temporary monopoly on—or profit off the production of, ACTs.[15] With malaria largely eradicated in the global north and with no major wars involving Western countries being waged in malarial areas, the disease had little impact on US or European populations. As a result, malaria was once again seen as a disease of the

poor, and Western pharmaceutical companies simply did not see an economic incentive to produce and market artemisinin-derived drugs.
Meanwhile, malaria cases continued to rise in many places in sub-
Saharan Africa. In the 1990s and early 2000s, the disease was estimated
to kill a child under the age of five every 30 seconds.[16]

Western pharmaceutical companies' lack of interest in producing
artemisinin drugs limited both its supply and its distribution. Given its
successful use in China, global aid agencies, including Doctors Without
Borders, began to put public pressure on the WHO in the 1990s to
endorse the drug. Aid agencies could not buy and supply drugs that were
not approved by the WHO and thus were unable to procure the drugs
directly from Chinese manufacturers. However, Novartis bought the
Chinese patent for a new ACT and made the drug commercially available
in 1999. In 2002, the WHO revised its recommendations and finally
endorsed ACTs as the first-line drug that should be given to malaria
patients.[17] While the initial price of Novartis's ACT kept the drug out of
reach for most people and public health agencies in the global south, public scrutiny forced Novartis to lower the price.

In negotiating the price drop with the WHO, however, Novartis protected its own corporate financial interests. In the 1990s and early 2000s,
the WHO was facing funding and legitimacy crises. Amid the broader
shift toward (neo)liberal economic policy and the emergence of the World
Bank as a major player in global public health, the WHO began to focus
less on public initiatives than on creating public-private partnerships.
Working along these lines, the WHO entered into a private agreement
with Novartis in 2002 that granted the drug manufacturer a de facto
monopoly on the production and sale of ACTs by preventing any United
Nations organization from sourcing the drug from other producers.[18] The
same year the WHO-Novartis agreement was signed, the Global Fund was
created to harness private capital to help finance global health initiatives.
The Global Fund had an explicit commitment to provide financial backing
for the WHO procurement of malaria drugs. The WHO's agreement with
Novartis and a steady stream of funding from the Global Fund thus created a structured economic market in which the company could sell its
discounted ACTs. In particular, the Global Fund provided the monetary
resources with which to purchase the drug, and the WHO's role in fore-

casting demand and distributing the drugs ensured that there would be a stable market for Novartis to profitably manufacture and sell its ACT in the global south.[19]

While Novartis benefited economically for several years from its de facto monopoly, its "corporate citizenship" image took a major hit. The Berne Declaration, an independent Swiss nonprofit organization (now known as Public Eye), that campaigned for more equitable relations between Switzerland and developing countries, began to lobby for the dismantlement of Novartis's monopoly in order to allow developing countries and other manufacturers to produce ACTs and distribute them through the WHO. In 2005, the WHO also started advocating for an end to its agreement with Novartis. The WHO admitted that current production levels of ACTs were insufficient to meet global demand and called for opening the ACT market to generic alternatives. According to the WHO, the increased production of quality generic ACTs would lead to lower prices through market competition and ultimately expand access to the drug for people that needed it most. Although Novartis's patents were technically valid until 2011, the company eventually recognized that defending its de facto monopoly would not be in its overall best interests. As a result, it decided to allow other companies to make generic ACTs starting in 2005. For several years following, Novartis still dominated the ACT market. The company held 85% of the total market in 2008, but by 2013 its overall share had dropped to 12%. Today, pharmaceutical manufacturers based in India, China, Germany, France, and the United States make generic versions of Coartem that are sold directly to individuals who can afford it in places like sub-Saharan Africa.[20] While a number of manufacturers now produce ACTs, Novartis still remains the key player in the sale of ACTs to *public* entities, thanks largely to a legacy of its partnership agreement with the WHO. In 2020, Novartis's drug enjoyed about 85% of the public sector market for ACTs. In addition, it is one of only five ACT brands officially endorsed by the WHO.[21]

Although ACTs have been commercially available since 1999, Novartis's monopoly heavily restricted its initial use. From 1999 to 2004, more than 95% of children with malaria in sub-Saharan Africa were treated with old and ineffective drugs like chloroquine. As late as 2010, long after Coartem became more widely available, it still cost several times more than chloroquine on the open market.[22] Some have claimed that the delays in getting ACTs to people in places like sub-Saharan Africa and the prohibitively high market price for the drug amount to genocide. Few public health systems in high-burden malaria countries could afford to purchase the new drug without outside support. Thus, many governments have had to rely on aid agencies based in Western countries, such as the Global Fund, to provide financial backing for drug procurement. In addition, most countries depend on World Bank or IMF loans to help them access ACTs and other essential malaria medicines.[23]

While poorer countries in sub-Saharan Africa suffer the most from malaria, countries in the global north have long driven antimalarial research, invention, and patenting.[24] More than 3,000 Artemisinin-related patents have been filed since 2000. The United States and China hold most of them, each with around 1,300 patents, followed by Germany, France, and India, with around 300 each. While the manufacturing of ACTs has spread to a larger number of countries, production of ACTs remains low or nonexistent in those places most affected by malaria, including Uganda.[25] Lack of technological know-how, poor access to education, and inadequate government funding are the factors most frequently purported to have inhibited the development of ACTs and other needed drugs in sub-Saharan countries. However, trade-related intellectual property rights, upheld by the World Trade Organization, also protect patents on pharmaceuticals developed by well-financed companies in the global north and thus restrict their production and distribution in the global south.[26]

With the introduction of the World Bank and IMF structural adjustment policies (SAPs) that liberalized developing economies and encouraged foreign investment, some pharmaceutical companies did emerge in places like Uganda in the 1990s. However, SAPs prevent governments from enacting policies that could protect domestic markets for locally manufactured ACTs. Therefore, local producers are often unable to compete with pharmaceuticals manufactured by companies based in the global north. Moreover, infrastructural challenges largely prevent sub-Saharan drug manufacturers from meeting WHO production standards, so most continue to make only basic products, such as pain killers, simple antibiotics, and vitamins.[27] For combatting malaria, typically, they are able to produce only outdated drugs like chloroquine that are much less effective than ACTs. The pursuit of shareholder value, SAPs, and broad trends toward privatization have not only restricted the availability and affordability of ACTs but also shaped people's ability to get a proper malaria diagnosis and appropriate treatment.

## MALARIA AND CONDITIONS FOR HEALTH CARE IN UGANDA

Uganda is considered a high-burden malaria country. It carries the third-largest burden of malaria globally, behind only Nigeria and the Democratic Republic of the Congo.[28] During the decades that ACTs were delayed in reaching populations in sub-Saharan Africa, health care systems in many countries, including Uganda, were going through a massive transformation. In Uganda, the state had played a leading role in administering health services since the colonial period, when the "health services funded by the state were overwhelmingly geared towards the needs of the European and Asian community."[29] After independence in 1962, the government focused on improving the livelihoods of Ugandans. It launched of an array of ambitious programs to expand infrastructure and build hospitals and schools across the country.[30] At the time, the WHO aided poorer countries such as Uganda by providing essential medicines and quality care to populations that needed it most. Essential drugs were provided free of charge in many African countries and could be obtained by local residents at community health centers. While Ugandans experienced improvements in their health and well-being during this time, the country's military dictatorship in the 1970s derailed many of the relationships it held with external partners and development aid providers. In turn, the provision of public health services started to break down.[31]

Upon its return to democratic rule and subsequent acceptance of aid and advice from the IMF and World Bank, the Ugandan government reduced public funding for the provision of health care. As in Brazil and elsewhere in the developing world, the IMF and World Bank viewed health care as a service better provided by the private sector and promoted policies that facilitated its advancement. Uganda pursued one such policy by subcontracting health care services to the private sector. Under the guise of economic efficiency, the IMF and World Bank encouraged the Ugandan government to lease or sell public hospitals to private health providers.[32] But perhaps the policy with the widest effect among Ugandans was implementation of fee-for-service regimes. In the 1990s, user fees were introduced, purportedly to recover costs and to make up for lost public sector funding. However, people in Uganda were often unable to afford the fees and thus strongly opposed them. Creating a financial barrier to access, the user fees resulted in a decrease in health care utilization. In turn, distribution of quinine and chloroquine via public health centers dropped. This decrease was due not to a lack of demand but a lack of affordable access.[33]

The IMF and World Bank conditional loans further impacted health care in Uganda by undermining the domestic production of pharmaceuticals. In addition to privatizing the health care sector and implementing user fees, Uganda enacted the usual slate of IMF and World Bank recommendations in favor of market liberalization. The government liberalized its domestic capital markets, removed its price controls and trade protections, and devalued and floated its currency to encourage exports and foreign investment. Such changes decimated Uganda's domestic pharmaceutical sector as imported brand name and generic drugs more easily flowed into the Ugandan market and eventually displaced many domestically produced drugs distributed by formal health centers. By 1990, domestic production of pharmaceuticals had virtually stopped, and most Ugandan pharmaceutical companies had gone bankrupt.[34]

The IMF's and World Bank's policy prescriptions for improving health care in Uganda had a number of adverse consequences, reflecting the systemic problems common to many countries forced to accept conditional loans from international lending agencies in the 1980s and 1990s. These impacts included inadequate financing of health services, an inequitable distribution of services, and excessive dependence on donor funding.[35] Such factors have affected the provision of health care in all of Uganda, but rural communities have perhaps suffered the most. Living in areas deemed unprofitable by private health care providers, rural Ugandans depend on the public health care system and make up more than 80% of its users. While a successful popular movement resulted in the end of user fees in the early 2000s, the public health care system has continued to deteriorate due to insufficient funding. In 2000, annual public health expenditures in Uganda amounted to less than $14 per capita. In 2022, per capita public health expenditures were nearly the same.[36]

These challenges have created a two-tiered health system in Uganda. The private health care sector caters to the affluent and foreigners who can afford to pay out of pocket the exorbitant prices for quality care. The public health care system is underresourced and unable to provide adequate service to those in need.[37] Many public health centers in Uganda lack electricity, running water, and basic equipment and can employ only a few improperly trained staff, resulting in high patient-to–health worker ratios. Consistent underfunding of the public health care system has also led many health care workers and trained staff to leave the public sector and join the private sector. And, if provided the opportunity, some doctors have emigrated to become medical professionals in more affluent countries.[38]

The underfinanced and underprovisioned public health care system in Uganda has also hampered the country's efforts to treat malaria, particularly among those most in need. Public health care centers suffer from pervasive medicine "stock outs."[39] Even as ACTs have become more widely available on the global market, inadequate public funding has meant that public health facilities are unable to purchase enough medications to keep up with demand. In some years, inadequate supplies of ACTs were reported across the entire country, preventing poor populations in particular from being treated for malaria when they most needed it.[40] But for those who can afford it, another option exists. They can buy medication from the local drug shops. As the only place where people can reliably obtain antimalarials, drug shops have become an important provider of medication in Uganda's liberalized health care landscape.[41] However, local practitioners and community members recognize a number of problems with this form of malaria management.

## DRUG SHOPS AND THE FAILURES OF "TREATING FROM HOME"

In July 2018, Kelly sat in the pediatric ward of the Bududa Hospital. The smell was rank and the air felt still and stuffy. Tattered and torn screens hung ineffectively from the open windows. Mosquitoes could easily fly in to feast on the young children that frequented the ward. The children often lay listlessly, their bodies emaciated by the *Plasmodium* parasite. But, on this Wednesday afternoon, the ward was nearly empty, not because children were no longer sick with malaria but because the hospital had no medicines, no IV fluids, and no clean syringes. It was late in the month and the hospital had been in a stock-out for more than nine days.

Bududa Hospital is one of only twelve public health care facilities that serve the more than 230,000 residents of the district. It is the only level-four, inpatient facility in Bududa. The other eleven public health facilities are mostly out-patient clinics that diagnose and sometimes treat the most common and basic diseases in the region, including malaria. Bududa also has two private health care facilities. However, one of these closed in 2020. The other charges a fee of 40,000 Ugandan shillings (around $12) just to be seen, and likely more for testing or medication. These prices bar the typical household in Bududa from accessing private clinics. Bududa Hospital, like the other public facilities in the area, lacks clean running water and stable electricity. In a discussion group, doctors from the hospital reported that the facility typically operates with a 60-to-1 nurse-to-patient ratio. In addition, on the day Kelly met with the discussion group, the chief medical officer (CMO) did not make it to work even though there is always supposed to be a CMO present at a level-four facility.

Health workers in Uganda often express frustration with the conditions under which they are forced to operate. A nurse in Bududa went as far as to say, "It's not a priority for the government to provide the legitimate money to get enough health workers. And that is actually not a problem of Bududa only. You go to most of the health centers, most of the hospitals, government hospitals, they'll still be the same. So it's become a song and we're tired of singing it." The issue of brain drain comes up repeatedly in conversations with community members as well. As one local man commented: "Because [in Uganda] you . . . you do not have enough doctors, you don't have enough midwives, enough nurses, and then people are here saying we export [health workers]. And, it's actually people in government hospitals that are applying to be exported most. So, you see, when they leave, then all the hospitals become stagnant." Among the multiple infrastructural and personnel challenges, everyone seems to recognize the stock-outs as a key problem in preventing and controlling malaria in the region. "The medications are needed, but the drugs are not there," one health worker stated matter-of-factly. Conversations and observations at health centers around the district reveal that the stock-outs create a chain reaction of consequences and coping mechanisms that jeopardize people's health and the treatment of malaria. People with malaria can either wait for drugs to show up at their local public health clinic, buy medicine in the private marketplace at the local drug shop, or travel to a different public health clinic with the hope that it will have the drugs available.

In many public health care facilities, including those in the rural parts of Bududa, medicines like ACTs are delivered only in limited quantities about twice a month. In the absence of medications, clinical staff in Bududa can only diagnose malaria. If a test is positive, public health care personnel have been forced to advise people to come back to the health care center when the new supplies of ACTs are expected to be delivered. But such delays defeat the opportunity of an early diagnosis. The *Plasmodium* parasites continue to multiply in the human body, making treatment more difficult. In addition, returning home while waiting for drugs turns humans into more deadly disease hosts, from which mosquitoes can more easily transmit the malaria parasite to the rest of a person's family and throughout their community. For a patient diagnosed with malaria who can afford to pay for ACTs, the doctor writes a prescription or verbally tells them they need malaria treatment. The patient is then sent on their way. If the patient is physically suffering from severe malaria, they may be admitted to the facility or referred to the hospital for admission. A family member, usually the mother or female head of household, then goes to a nearby drug shop to get whatever medications or treatments are prescribed and brings them back to the hospital to be administered.

Many Ugandans cannot afford to purchase ACTs on the private market and know that waiting for them to arrive at the local clinic can be both dangerous and futile. Attempting to navigate the constant stock-outs, many community members talk about sharing information with their family and community about where medications are available at public clinics in the district. A local health worker described this strategy: "Actually, they first might go to the lower [public] health units, and then they find there are no medicines there. So, then, they send them by [word of] mouth, 'there is a stock out.' Or during times when the medicines are there, you will hear about it. 'Go to Bududa Hospital. It's where you will find treatments.'" Spreading the word that medicines are in stock, however, can lead to other problems, such as a flood of patients pouring in to a facility said to have stock. Regrettably, overcrowding also deters people from seeking care. As one doctor explained, "Now, for example, if a woman comes and sees a line of about two hundred, you begin to estimate the time it's going to take. And she'll leave, even if her child is very bad off." Long lines, lack of quality care, stock-outs, and the lengthy travel time to public facilities all discourage people from seeking health care through formal channels.

People suffering from malaria or taking care of ill children are thus forced to choose. They can walk—sometimes for miles—to a public clinic that is likely to be overcrowded or out of medicine, or they can go to the nearby drug shop and purchase antimalaria medicines in minutes. Community members and health workers alike report the common perception that visiting a public clinic is a waste of one's time. People know that when they visit a public facility, they are likely just to get an official diagnosis. As one community member put it simply, "But like [going to] the hospital's all about going to just test, and then writing you the medicine you're supposed to get, and you buy the medicine yourself."

There are many drug shops in Bududa, even in the more remote areas. For those who can afford it, the decision is thus often easy. As a community member plainly stated, "Before you go to the health center, you can treat malaria from home. You can go and get medicines from the drug shop. That is what people do." From a public health standpoint, the main advantage of going to a health facility in Uganda is the diagnostic test that public and private hospitals and clinics are required to administer before prescribing or providing treatment for malaria. But the drug shops in areas such as Bududa sell malaria medications to anyone willing to pay. Neither a prescription nor any official form of proof that a person has malaria is required. Some drug shops in Uganda do offer rapid diagnostic testing for malaria, but customers must pay an additional fee for this service and therefore typically do not get a test.

While lower-cost generic ACTs can now be easily found in most of the drug shops that have proliferated in the Bududa District, a significant portion of the population is still unable to afford them. Community members frequently comment that ACTs are "expensive." Numerous households describe having to sell livestock or other items to "get money for treatment." Others say they are simply "not able to get—to pay for treatments. So they suffer." And many people in Bududa report that some malaria drugs, like quinine and chloroquine, are cheaper and easier to find in the pharmacies. They are thus sometimes used first by residents who cannot afford ACTs. "The 'new drug' [ACT] is most expensive," some community members say.

People in Bududa are sometimes forced to buy half doses or child-sized doses of an ACT because they do not have enough money to buy the proper regimen. Many people also stop using the medication partway through the course. They think that stopping treatment will help them to avoid economic hardship in the future. Health care workers discussed this challenge at length. One nursing officer reported, "We find that they are not taking all the medication. They take some, and when they feel, like, better, they leave it. . . . So, so many halve the treatment." By saving some pills for the next time when someone in their household falls ill with malaria, which is likely only weeks if not days away, they think they will be able to treat multiple people with the number of pills necessary for one person to safely rid themselves of the *Plasmodium* parasite.

The challenges that result from taking partial doses was a prevalent theme in the discussions Kelly had with health care workers. Many emphasized that self-treating from home can complicate the illness and lead to undetected severe cases of malaria. As one worker explained: "Yeah, because what happens is someone will be presenting with severe malaria and they've [been] subjected to a lower dose at home. So, when they come here [to the public hospital], it means usually we do a test and if you've mismanaged the process before, they might test negative. But, by the time a patient comes here, he's really wasted, he's really down. He might not recover." Patients who develop severe malaria are much harder to treat. They often require intravenous treatments and lengthy hospital stays, putting major strain on households and caregivers. The underfunded public facilities in Uganda lack even mattresses, blankets, food, and drinking water for patients. All of these items must be provided by families, and family members must remain at the hospital to assist in caring for the sick. A child or adult patient who recovers from severe malaria often deals with lingering, even lifelong after-effects, including permanent damage to the brain or other vital organs. People in Bududa often link mental illness with suffering from severe malaria, observing that those who survived a complicated malaria case often have permanent neurological impairment.

Antimalarial drug resistance is also a common theme when Ugandan health workers reflect on issues arising from "treating from home". As one clinical officer working in Bududa eloquently put it:

> And somehow, as soon as the child starts improving, they discontinue the use. And that is why we are having problems with resistance. With malaria we say, Why? Why? Western drugs come and after two years, and we're resisting? So misuse is the main problem. People take half doses. Some of them take the medications; you have a headache and you say I have malaria—you take antimalarials. I think that needs a diagnosis. When they take the medication, they need to know for what condition.

Especially in areas such as Bududa where malaria is endemic and common, many community members perceive that they can easily recognize the symptoms of malaria in themselves or in their children. While malaria is one of the most widespread illnesses in the district, it shares symptoms with other common health threats in the region, such as flu and different infections resulting from contaminated food or water. But taking ACTs or other antimalaria drugs when one does not have malaria can lower the drug's effectiveness when it is needed shortly thereafter.

Despite the prominence and salience of drug shops in Uganda, the regulation of these providers is generally low. In principle, drug shops in Uganda must be licensed, and owners are required to have some formal medical or pharmaceutical training. However, the dearth of public funding stemming from the implementation of IMF and World Bank SAPs has reduced the Ugandan government's capacity to regulate key industries, including various components of the health care sector. In practice, the formal requirements for registering drug shops are often not enforced. Drug shops are often staffed with undertrained personnel, may stock outdated or inappropriate medications, and have generally come to represent a lower-quality outlet for health services in Uganda.

The perspectives of community members and health workers in Bududa align with a growing body of research from across Uganda and other countries in sub-Saharan Africa which illustrates that drug shops typically represent the first step in seeking treatment for malaria.[42] One report based on household survey data that examined the practices surrounding the use of drug shops found that over 80% of the population surveyed typically first "treated from home" before spending time to visit a public health care facility and get a diagnosis. The consequences of these coping strategies for public health are numerous. The issues surrounding stock-outs and needing to treat from home lead to delays in seeking medical care and a proper diagnosis, facilitate the mismanagement of medications, and enable improper dosing. Many commented that the medications are not as effective as they used to be. One woman tearfully observed while recalling her sister who died of malaria in the year prior. "You can take the medicine, but you still die." Others stated that malaria "never seems to be cured" anymore and that they have frequently recurring cases, something that did not used to happen in the past. Not only are complications and the risk of death higher after antimalaria medicines have been mismanaged, but these trends also pose major challenges for disease salience. The fact that most people in Uganda do not visit a formal health facility when they think they have malaria means that the vast majority of malaria cases are likely missed in official reporting systems. The prevalence of malaria in many districts is thus likely to be much higher than the official statistics would suggest.

## FINANCE, MALARIA, AND ACT DERIVATIVES

When local people in Uganda are asked about how to address the ramifications of treating from home, they often give a simple response. As one health worker said candidly, "The solution is there: the government should provide full doses." Another noted, "Yes, if [the government] provided the medicine, they would take [the ACTs] and not abuse it." Oddly enough, despite confronting challenges of a fragmented and privatized system every day, health workers in Uganda do not dismiss or decry drug shops and their owners. Kelly could not help but ask a health worker she has known for years, "Is there friction—friction between these private drug shops and the public clinics, especially when you can point out all the ways they make it hard for you to treat malaria effectively in the community?"

He gave the question some thought before responding, then said, "Yes, I would only think that there is, there is friction, look at all the mishandling [of patients] that happens. But even so, during the stock-outs they are the people who do help us."

Drug shops have thus become a necessary and even valued part of the privatized health care system in Uganda. Without them, there would be no way for people to get the medications they need. However, conditions placed on loans to Uganda negatively affected health care in the country and generally reduced the ability of the poor and vulnerable to access treatment for malaria. Not all households can afford to purchase ACTs, and halving of doses as a coping mechanism will only compromise the effectiveness of the drugs and breed resistance over the long run.

In 2015, artemisinin-resistant *P. falciparum* was officially identified across several countries.[43] According to Eloy Rodriguez, an endowed professor and research scientist at Cornell University, "[*Plasmodium*] is in a battle, and it wants to live too. It doesn't want to die, so resistance is going to be around forever."[44] Ironically, climate change and environmental degradation are also compromising the ability to grow *Artemisia annua*, the tree that ACTs are derived from. Currently, China and Vietnam have accounted for about 80% of the harvest volume of *A. annua*, while East Africa has accounted for the remaining 20%. It takes eight months' minimum growth time for the tree to be harvested. After extraction, artemisinin is sent to specialized manufacturers in the United States, Europe, China, or India to be converted into ACTs. The entire process, from the planting of the seed to the finished product, takes about fourteen months. These long lead times for manufacturing ACTs are likely to affect availability if growing conditions continue to worsen.

While there has been an uptick in patents involving *A. annua*, especially since the Nobel Prize was awarded in 2015, current research is not focused on creating new malaria medications. It is more economically prudent for pharmaceutical companies to find new uses or applications for existing derivatives than to develop new treatments. The potential ill-fated use of chloroquine to treat COVID-19 is a case in point, but the same is happening with ACTs. The patents being filed currently relate to *A. annua*–based treatments for chronic diseases and certain cancers.[45] In a for-profit health system, there will always be a void in research and funding for conditions that largely affect only poor people.

These trends are reflected in the latest malaria intervention to make headlines—a potential vaccine. The parallels between the development and rollout of ACTs and the new malaria vaccine are striking. The core ingredient of the pathbreaking malaria vaccine was actually discovered about thirty-five years ago, and researchers have known since at least the late 1990s that the formula was fairly effective at protecting against malaria.[46] It was not until 2016 that the WHO recommended further evaluation of the vaccine, called RTS,S or Mosquirix, and it took more than four years for the first trials to be conducted in a few high-burden malaria countries.[47] As in the case of ACTs, the disparity between the time the intervention was discovered and developed versus the time it took to be administered to the people who need it reflects the priorities of financed capital.[48] The interests of profit making facilitate long-standing and structured patterns in which deadly diseases that afflict people do not matter. As Ashley Birkett, director of the malaria vaccine initiative at the non-profit global health organization PATH, observes, the people who are affected by malaria "[are] not Europeans; they're not Australians; they are poor African children. . . . Unfortunately, I think we have to accept that that is part of the reason for the lack of urgency in the community."

Funding deficits have also plagued every step of the malaria vaccine development process, and experts already worry that funding shortfalls will hamper the rollout of RTS,S.[49] Funding for research and development has been on a downward trend since 2017. In 2020, funding for malaria vaccine research dropped further as the search for a vaccine and cure for COVID-19 took first priority and promised to be more profitable. The full course of the malaria vaccine will cost an estimated $38 to administer per child. By comparison, vaccines for diseases such as polio are available for less than $1. Thus, a potentially life-saving malaria vaccine is available but is unlikely to reach the arms of the most vulnerable children anytime soon. Beyond financing there are other concerns; for example, the malaria vaccine may be only partially effective, and it requires multiple doses that must be given at certain times during the first two years of the child's life.[50] The wide-scale roll out of the malaria vaccine thus clearly faces many challenges.

For now, ACTs are the best means of combating malaria in places like Uganda. However, a health system motivated by the interests of finance capital facilitates misuse and an uneven use of these medications, ultimately compromising their effectiveness. Resistance to ACTs is already developing and spreading.[51] Local Ugandans notice this shift and link it to an increase in malaria cases and complications in their communities. Ultimately, finance capitalism leads some human lives to be valued more than others. Those who are unable to afford malaria medications or live in places where they cannot be easily accessed are deemed expendable. While patents and liberalized economies are prophesied to simultaneously spur shareholder value and local economic development, they have often led to the dismantling of public health systems, created a void of investment in eliminating low-return diseases, and compromised the ability of people, especially the poor, to lead healthy and productive lives. And, in this world created by finance capital, the malaria parasite has spread, become more drug resistant, and thrived.

# 7  The Bank of the Planet

The words *pathogens* and *finance* do not frequently appear together. Pathogens are most often perceived of in a negative light and not thought of unless causing harm. Most people, at least in the global north, are reminded of the existence of pathogens only when they catch a cold, the flu, or perhaps COVID-19. In the mainstream media, pathogens become more than a human-interest story only when an infectious disease outbreak has the potential to cause debilitating illness or death in more affluent populations. Finance, on the other hand, is conceptualized in the popular imaginary largely as a necessary component of capitalist society. While the reputation of finance has not exactly been glowing in recent decades, many of us are invested, or told we should be, in its so called success. We can thus easily track finance by reading a section dedicated to its existence in most major newspapers, by listening to or watching a broadcast dedicated to monitoring its trends, or by simply opening an app. In such news sources, finance is rarely painted as if it were as bad as a pathogen, and it is infrequently, if ever, linked to causing physical human harm or death.

However, recognition that financialization has caused an array of social harms has led some mainstream media columnists and popular fiction writers to question the role of finance and financiers in contemporary society. Referring to financiers as rentiers, Paul Krugman argues that we are basically living under the "rule of rentiers," whose interests are continually satiated by policy makers at the expense of everyone and everything else.[1] Going a step further, Martin Wolf, the chief economics commentator at the *Financial Times,* blames finance capitalism for stagnating economic growth and incomes and proposes ways to "wipe out rentiers."[2] And in his popular novel *The Ministry for the Future,* Kim Stanley Robinson depicted the financial class as a "parasite killing its host by overindulgence."[3]

Such critiques of finance and financiers have long been present in academic circles and can be found in the work of many early social theorists. Adam Smith and Karl Marx alike depicted financiers in a pejorative manner, seeing them as nothing more than a class of people extracting wealth from those who actually produce things. Perhaps not so coincidently, illness and death from infectious disease outbreaks such as smallpox, yellow fever, and cholera were common in the societies on which such theorists based their observations of financiers. More recent critiques have emerged as global policy shifts over the past fifty years have revitalized a financial class and the associated forms of finance capitalism.[4] And, during this time, illness and death from infectious diseases such as Lyme, dengue, Zika, and malaria have regularly plagued societies.

The latest round of finance capitalism emerged with the end of the Bretton Woods Agreement in the 1970s, which unleashed global capital and allowed it to flow across borders more easily. A global debt crisis followed in which a greater number of countries were forced to allow such capital to flow into their economies with fewer strings attached. As this process occurred, both the global landscape and global public health were turned into financial assets. Such changes frequently increased the shareholder value of corporations in the global north and helped create an array of new financial products that extracted rents from society and nature and placed them in the hands of Wall Street financiers. However, the newly created financialized landscapes also proved to be more conducive to the occurrence and spread of infectious diseases. In addition, the creators of the purported cures to such infectious diseases under the auspices of global public health often prioritized the creation of financial profits over

people's health. But if the spread of infectious diseases and their inadequate treatment can be attributed to contemporary finance and finance
capitalism, what is to be done?

## A NEW FINANCIAL LANDSCAPE

It could be argued that dismantling the current global financial architecture and rebuilding the Bretton Woods institutions to resemble their mid-
twentieth century form would reverse the trends and alleviate the burdens
associated with the increase in infectious disease outbreaks. Capital controls once slowed global financial flows and provided countries with
greater control over their own currencies and economies. Tariffs and other
trade barriers allowed domestic industries and agricultural producers to
earn more sustainable incomes. And social democratic and developmental
state policies sought to foster the livelihoods and well-being of people by
more generously funding the expansion of infrastructure, the provision of
health care, and a panoply of other government programs and social safety
nets. While such measures previously did, and likely would again, reduce
infectious disease outbreaks and lessen their overall impact, they still
privileged the hegemonic power of the United States and worked to
advance and expand the interests of private capital. More than likely, creating a global financial architecture that truly works to promote the health
of humans and the planet requires something new.

   One way to privilege people and public health over the profits of corporations and their shareholders is to democratize finance. The democratization of finance can mean many things. In a basic sense, it entails "the
opening of financial opportunities to everyone."[5] Implementation could
involve providing access to bank accounts or extending credit and investment opportunities to the poor and marginalized segments of society that
have often been excluded from formal banking and financial institutions.
Both programs have been tried in various parts of the world. However,
merely providing credit and investment options does not guarantee an
optimal outcome. As others have demonstrated, providing a loan to a non-
traditional borrower may enable a person to start a business or survive an
economic downturn, but depending on the conditions of the loan, it could

also entrap them in endless cycles of borrowing and, in turn, deepen their poverty. At the same time, extending access to loans, bank accounts, and other financial instruments does not actually ensure financing is used in productive and progressive ways.[6] Plans for the democratization of finance that focus on individual access to and participation in banking and financial markets do little to alter the politics and power relations in the existing financial world, most of which function to increase the wealth of the already well off at the expense of everyone else.[7] With financial wealth currently extracted at the expense largely of the poor and the planet, merely enhancing access to finance would probably do little to decrease infectious disease outbreaks or improve our collective responses to them.

What is thus needed is a way to weaken the existing dominance of the financial system by the extremely wealthy and a handful of giant private financial firms. This entails not only offering banking, lending, and investment opportunities to those who have traditionally been excluded, but also changing who controls and benefits from such opportunities. Existing public institutions could be allowed to exercise greater power over private finance and the broader financial system. The ability of private finance to operate and profit is granted and underwritten by sovereign public entities. In the United States, the, the, the Federal Reserve, the Federal Deposit and Insurance Corporation (FDIC), and the Securities and Exchange Commission (SEC) all regulate banks. In addition, the Federal Reserve and the Treasury Department ultimately control how much money and credit is flowing by selling public liabilities (such as Federal Reserve notes and Treasury securities), buying privately issued debt, and printing and circulating money.[8] As social theorist Karl Polanyi long ago pointed out, money—in this case finance—is not dictated by the purported laws of supply and demand.[9] Money is a fictitious commodity created and governed by society. For much of the past fifty years, the institutions responsible for regulating finance have increasingly accommodated the profits of private finance over the public good. It is technically in such institutions' regulatory capacities, however, to accommodate the latter over the former.

Over the past few years, government interventions in finance have provided an example of how this accommodation could occur. COVID-19 demonstrated to even the most casual observer some of the ways the financial sector, human health, and the environment are linked. The initial outbreak of the virus prompted the US Federal Reserve and other central banks across the globe to make unprecedented interventions in the financial marketplace. To stave off the collapse of the global financial system, central banks dropped interest rates to near zero and aggressively purchased securities to bolster the supply of available money. The Federal Reserve also eased access to dollars globally by opening an array of temporary swap lines with other central banks.[10] At the same time, governments put into place—albeit for a limited time—an array of progressive policies and programs that showed what is possible. While insufficient in amount, direct payments to families that helped them through the pandemic provided a view of what a universal basic income could do to decrease poverty. Universal school meals in public schools showed how childhood hunger and nutritional deficits could be potentially addressed. And access to free vaccines and testing demonstrated how public health care could work in places where it did not already exist.[11] However, by 2022 the threat of global financial collapse had receded, and policy making returned to largely protecting the interests of the elite. The most progressive policies and programs implemented in response to COVID-19 have nearly all been ended, and the regulators of global finance have once again sought to secure the wealth of financiers at the expense of everyone and everything else. The long-term prioritization of the public good over finance thus likely requires more than a pandemic- or a disaster-induced response.

What is necessary is the creation and institutionalization of countervailing forms of power to challenge the dominance that the wealthy and large private investment firms hold over finance. Some have suggested building more robust public and nonprofit financial institutions.[12] For example, credit unions could be promoted as an alternative to mainstream finance. Such institutions have a history of localized control and investments. Each member-depositor usually has a vote on who serves in leadership positions. Credit unions have also traditionally sought to loan money within the communities they serve. In addition, as nonprofits, credit unions are driven not to achieve profits for shareholders but to respond to the interests of members. Such efforts could be built upon by providing credit union members or their broader communities with more direct say over what gets financed. For example, credit union members could seek to collectively decide how such institutions loan money, refusing to finance projects that are harmful to humans and the environment while privileging socially beneficial projects and businesses. Workers with pensions and nations with sovereign wealth funds could also weaponize their assets in a similar manner.

But the democratization of finance would have to occur beyond the local and even national scale. Global flows of finance currently favor financiers and the financial class, particularly those in the global north. The dominance of the US dollar as the global currency also subjects foreign countries to the whims of US domestic and foreign policy. As a result, the democratization of *global* finance would likely require changes to global financial flows, the decentering of the US dollar as the dominant currency used in trade and finance, and the creation of new global financial institutions or the reconfiguration of existing ones. With the dogma of free trade and the power of the World Trade Organization having faltered over the past decade, it has become increasingly possible that countries may turn to restricting or regulating global financial flows into their national economies, limiting short-term speculative investments, and enacting protectionary measures to insulate strategic or fledgling economic sectors from undue competition.[13] At the same time, the power and dominance of the US dollar has come into question. The rise of China as a global hegemonic challenger to the United States has elevated the use of the renminbi in international lending and trade, providing some nations with an alternative to the U.S. dollar. Alternative trading currencies have also been proposed elsewhere, such as the sucre in Latin America. In addition, alternative international development banks, such as the BRICS banks, have emerged over the past decade to serve as potential alternatives to the IMF and World Bank.[14]

There is no guarantee that the democratization of finance would lower infectious disease outbreaks or improve how they are treated. However, fostering the aforementioned institutions offers the possibility of using finance to prioritize the health of humans and the planet over the profits of private corporations and their shareholders. Providing people, communities, and nation-states both with greater access to finance and greater say over how it is used could result in lending and investment practices that create landscapes less susceptible to infectious diseases and health care systems better able to treat them.

## THE POTENTIAL POLITICAL ECOLOGIES
## OF DEMOCRATIZED FINANCE

To democratize finance and use it to decrease the likelihood of infectious disease outbreaks, what exactly would need to be done? In particular, how might finance be used to create ecologies that could lower incidence of infectious diseases in humans? In a basic sense, finance would have to be used for public good. In some ways, this already occurs. Banks offer loans, bonds, and other financial instruments to public entities. Such products are packaged and sold in the financial marketplace with lower interest rates but with the expectation that they will yield stable returns. However, these are not necessarily investments targeted to addressing infectious disease.

Perhaps the most prevalent means of landscape change known to decrease infectious diseases has been the construction, improvement, and maintenance of water, sewage, and sanitation infrastructure. Tainted water can hold a variety of bacteria known to wreak havoc on the human immune system. In addition, inadequate water delivery and removal as well as lack of sewage and sanitation services are known to proliferate pools of standing water, ideal places for insects such as mosquitoes to reproduce and carry diseases such as malaria, dengue, and Zika to humans. With existing infrastructure in many places aging past its projected lifespan, and some places across the globe never having built such basic infrastructure, finance could be directed toward providing everyone with access to clean water supplies and adequate sewage and sanitation services.

For much of the past century, federal, state, and local governments funded the expansion of such critical infrastructure. While some of this funding came from tax dollar savings, the up-front costs of infrastructure often required funding via loans, bonds, or other financial mechanisms. If seen as stable economic entities, many governments can obtain long-term loans and issue bonds with relatively low interest rates to fund infrastructure projects. However, access to infrastructural finance has historically been far from equal. In the post–World War II era, the heyday of the US infrastructure build-out, Whiter and wealthier communities received better rates on loans and bonds than minority communities, and the latter often paid more for credit or lacked access altogether. Similar inequalities existed in global finance, with countries in the global south largely unable to borrow at low rates or issue bonds that had more than marginal returns.[15]

Infrastructure loans and bonds are thus nothing new, but a system of democratized finance could design and offer such financial instruments with low interest rates to all or even prioritize places most in need. Although credit unions with a local focus would likely lack the capital necessary to back such loans, they could act collectively to pool assets and more actively partake in the sale of loans and bonds on the secondary market to provide such financial services on a small scale. Alternative development banks backed by the assets and credit ratings of several nation-states could do the same and likely operate on a larger scale. Another potential alternative would be for governments or banks to set up a system for issuing loans and bonds similar to progressive forms of taxation. Even though systems of lending have traditionally been set up in reverse, governments or public banks could institute policies in which the wealthy pay marginally higher interest rates to essentially subsidize the borrowing of those less well off. Such a system of lending would undeniably be difficult to implement, but an argument could be made that preferentially financing infrastructure in lower-income places—where many infectious disease outbreaks tend to originate—is good for everyone. After all, insect vectors and infectious diseases know no borders and do not differentiate between the blood and bodies of the rich and those of the poor.

But funding critical infrastructure is not the only way finance can be used to abate infectious diseases. While infectious diseases are more likely to plague underserved densely populated areas that lack adequate infrastructure, they often emerge in recently disturbed and altered ecosystems. Landscape change can make places more conducive to the presence of infectious disease hosts and vectors, whether due to the greater presence of a food source, the creation of a site of reproduction, or the absence of viable competitors and predators. For example, logging can result in newly created forest edge where certain types of plants and the herbivores who eat them proliferate and potentially serve as an infectious disease hosts. Or a clear-cut open field can contain a plethora of standing water where insect vectors such as mosquitoes thrive and the population of predators such as birds is too low to decrease their populations. Ecological distur-bances are often the result of human activities, such as the expansion of human settlement or the intensification of a particular land use. As this occurs, humans come into more frequent contact with animals or insect vectors that could carry infectious diseases which previously rarely spilled over to human populations. However, if the ecology of infectious diseases was accounted for in finance, loans and investments could privilege the continual regeneration of biodiverse landscapes.

Over the past decade, alternative forms and sources of finance have been developed under the auspices of making environmental protection profitable and challenging the hegemony of Wall Street. Investment firms have devised a number of financial opportunities for those seeking to put their money where it can purportedly "do good" for the environment. Perhaps the best-known example has been labeled environmental, social, and governance (ESG) investing. Through such investing, everyday peo-ple can invest their money in mutual funds composed of companies sup-posedly engaging in more sustainable business behavior. Progressive observers have noted, however, that such funds and sustainable investing more broadly rarely uphold the promises and values they seemingly pro-mote.[16] At the same time, alternative forms and flows of global finance have not radically departed from the past. Chinese loans—many of them made in renminbi—have flowed in increasing amounts throughout the world but most have gone toward funding extractive industries and trans-port infrastructure, activities known to drive massive landscape change. In addition, the terms of the loans have often been no better and perhaps even worse than some of those offered by the IMF and World Bank during the neoliberal era.[17] The contemporary changes to global finance have thus often left places in environmental ruin and heavily indebted, two fac-tors known to increase the occurrence and severity of infectious disease outbreaks.

Besides using democratized forms of finance, critical infrastructure projects could privilege ecological regeneration over ecological destruc-tion of existing landscapes when loans are made and financial instruments are developed. The driving criterion for such loans or investments would not be how quickly they can turn a profit but how they affect the environ-ments in which we live. Enforcing this standard could be done with carrots or sticks. To incentivize the creation of biodiverse landscapes, more favorable interest rates could be given to projects and places that actively seek to regenerate biodiversity and pursue ecological solutions to prevent infectious diseases. At the same time, banks and financial firms could outright reject loans for and pull investments from activities known to harm the environment and foster infectious diseases. Given that conventional banks and international lenders have proven incapable or unwilling to meaningfully account for the environmental consequences of their financial activities, alternative banks and interventions would likely be necessary to make such policies or programs possible.

The democratization of finance in creating landscapes less conducive to infectious disease outbreaks would need to account for the local economies and ecologies in which the financing was made available. In each of the cases discussed in this book, the necessary financial mechanisms would vary. In the United States, democratizing mortgage finance could result in the creation of landscapes with fewer ticks and tick-borne diseases. In Brazil, democratizing global finance through the creation of viable alternatives to international lenders such as the World Bank as funders of large-scale infrastructure and other development projects could prevent Brazil from falling into a recurring debt trap, help the country receive better interest rates on loans, and in turn install the infrastructure necessary to prevent future outbreaks of dengue, Zika, and other mosquitoborne diseases. And in Uganda, alternative international lenders, ones that privilege biodiversity, could help small coffee growers stabilize their incomes by promoting and selling quality, shade-grown coffee in smallscale enterprises over fast, sun-grown coffee and, in turn, potentially create less ideal environments for mosquitoes to transmit infectious diseases such as malaria.

Accounting for the ecology of Lyme disease in mortgage financing in the United States could decrease disease incidences in a number of ways. First, mortgages could be regulated so as to encourage the downsizing of the American residential lot and raise housing density. Such a policy may seem impossible, but for much of the postwar era, Federal Housing Finance Agency (FHFA) loan limits constrained the size of lots and houses. Whereas the FHFA purchased mortgages in most of the United States with a value up to US$726,200 in 2023, in 1970 the agency purchased only mortgages with a value up to US$30,000, or approximately $US260,000 when adjusted for inflation.[18] Lower FHFA loan limits would not prohibit the construction of large homes on large lots, but they would make such purchases more difficult. Encouraging denser forms of residential construction, more space in local communities could be preserved as contiguous greenspace where the presence of more biodiverse species would be likely. Second, finance could be used to maintain and create more biodiverse landscapes. Most conventional residential construction results in the complete clearing of a landscape before building commences, and then the finished product is often surrounded by a sea of turfgrass.[19] Such a process heavily disturbs the existing ecology—whether forests, grasslands, or something else—and replaces it largely with a single nonnative species. While the dominant ecology of the single-family home may be conducive to a game of croquet and provide a burgeoning herd of white-tailed deer with a steady food supply, the ecological balance needed to keep the population of the latter under control is lacking. Taking these facts into account, finance could perhaps shape the ecologies of constructed residential areas by providing builders with lower interest rates on land purchased for future homes if they agreed to construct residences within and as part of a locally biodiverse landscape. Developers could, for example, maintain an existing forest or grassland as part of single-family homes' yards, or use native species to landscape a residential lot built on a previously altered ecology. In both denser communities with larger adjacent contiguous greenspaces and communities with residential landscapes planted with local native species working to enhance biodiversity, it is possible that the natural predators of Lyme disease hosts and vectors would increase and help to lower the rates of Lyme disease in human populations.

Incorporating the ecology of dengue and Zika in global finance could help deter outbreaks of diseases carried by the *Ae. aegypti* in Brazil. Global lenders could be pitted against one another so that Brazil receives more favorable terms of trade and finance. During the Cold War, countries sought to balance their relationships with the United States and the Soviet Union to extract better trade deals and aid from each. In the contemporary era, the potential exists for countries to follow this practice once again. Most of Brazil's debt is currently held by the United States and its allied international lending institutions, but as of 2022, Chinese development banks had loaned Brazil more than US$30 billion.[20] Chinese commercial banks, too, have established an increasing presence in Brazil over the past decade, not only by operating as lenders in Brazil but also by easing trade between the two countries by facilitating currency swaps.[21] The latter has allowed Brazil and China to circumvent Western banks in trade between the two countries and ever so slightly disrupt the power of the US dollar as the dominant currency of global trade. Luiz Inácio Lula da Silva, reelected as Brazilian president in 2022, has promoted efforts to further counter the US dollar in international trade and lending and has revived talk of a regional Latin American trade currency. Brazil and its states and municipalities are yet to use the contemporary shifts in global finance to shape the landscape in ways that could curb mosquito-borne diseases. Loans from the World Bank currently fund more water, sewage, and sanitation infrastructure than do those from China.[22] In addition, a significant portion of the loans from China have gone toward expanding extensive agriculture in the Brazilian interior, which contributes to the increasing incidences of *Ae. aegypti*–borne diseases in the country.[23] However, with the United States and China locked in an intensifying geopolitical rivalry, Brazil could seek to obtain better terms on international loans for water, sewage, and sanitation infrastructure projects or try securing loans that it could more freely use at it chooses. At the same time, obtaining loans not in US dollars could help Brazil somewhat better insulate itself from fluctuations in US monetary policy. The latter strategy could allow Brazil to continue funding infrastructure projects without having to worry as much about fluctuating interest and currency exchange costs.

In Uganda, better accounting for the ecology of malaria in global finance could help deter outbreaks of the disease in a number of ways. As in Brazil, Uganda could attempt to secure more favorable terms of trade and finance by taking advantage of the geopolitical rivalry between the United States and China to secure lower interest rates and decenter the US dollar as the dominant trade currency. Over the past decade, Uganda has used Chinese-backed financing to fund an array of infrastructural projects, but often on terms no better than those of the IMF and World Bank loans it took on in the 1980s.[24] To best employ global finance to decrease the likelihood of malaria, it would probably involve funding sustainable forms of export agriculture that prioritize quality over quantity. For coffee producers in Bududa, this funding could entail using aid or loan moneys to reform and revive coffee cooperatives. Cooperatives function in ways that collectivize financial risks and rewards. For financiers, cooperatives may be seen as less risky recipients of loans. One farmer's bad harvest would not necessarily mean a default on payments. Cooperatives could also use their collective power to bargain for better interest rates and better terms of trade. Furthermore, investors could help build the power and self-sufficiency of cooperatives by potentially helping them to create their own banks or to finance ways to add and capture greater value from the coffee local farmers grow. Investors could assist them in restarting local coffee bean–hulling facilities or fund local roasting activities so that fully processed coffee could be exported and sold at a higher price. Ideally, farmers would thus utilize more sustainable cultivation techniques, such as shade-grown coffee, and the strategy could potentially decrease mosquito populations and human exposure to the diseases they can carry.

## THE HEALING POWERS OF PUBLIC GOODS

Using finance to create more biodiverse landscapes could curtail infectious diseases but would not eliminate their occurrence. Humans will always encounter animals and insects that carry infectious diseases. In addition, infectious disease hosts, vectors, and the bacteria, viruses, and protozoa that cause them will evolve and eventually adapt to live in new environments. Communities and individuals will thus still need to take steps to address infectious disease outbreaks when and after they occur. Taking these facts into account, how might democratized finance be used to diminish the effects of infectious diseases in humans?

In the most basic sense, improving access to health care would blunt the negative impact of infectious diseases in human populations. With a proper diagnosis, bacterial infections such as cholera, tuberculosis, and the plague can usually be treated and cured with antibiotics. Similarly, parasitic infections such leishmaniasis, babesiosis, and malaria can usually be treated and cured with antiparasitic drugs. Viral infections such as yellow fever and dengue have less specific treatment options, but for those with severe cases, having access to hospital care can be the difference between life and death. In addition, an array of vaccines have been developed to prevent or decrease the impact of infectious diseases such as yellow fever and polio. Finance could thus be used to enhance the prevention, diagnosis, and treatment of infectious diseases by bolstering access to conventional means of medicine and health care.

But as infectious diseases outbreaks have increased over the past three decades, finance has been held up as a driver of innovation in health and medicine. Venture capitalists and large pharmaceutical companies, backed by their shareholders, seem to have provided the world with vaccines and new treatments for an array of infectious diseases, perhaps most notably COVID-19.[25] However, this perception overlooks the fact that basic research for the science and technology underpinning many vaccines and treatments for infectious diseases either originates in public institutions or is publicly funded.[26] For example, the mRNA technology used by Pfizer and Moderna for their COVID-19 vaccines was developed and advanced over fifty years by scientists at public universities funded with public grant money.[27] At the same time, access to vaccines, medicines, and other forms of medical treatment is far from equal across the globe. Pharmaceutical companies' drive to create profits for shareholders often makes the price of vaccines and other medications unaffordable. In addition, as has been demonstrated with the development of mRNA-based COVID-19 vaccines, poorer countries deemed to have unprofitable market potential have frequently been excluded from vaccine and other medical supply chains.[28]

Given that public funding already supports much of the basic research necessary for pharmaceutical innovation, democratizing finance could entail a redistribution of who benefits from advances in health care. Such a system could turn public funders into shareholders. For example, in the United States, public institutions such as the National Institute of Health or the National Science Foundation could become shareholders in the pharmaceutical innovations for which they provide initial funding. Most proposed innovations fail or advance knowledge only marginally, but publicly funded grants could be seen as venture capital, and viable products that make it to the marketplace could provide the public with a return on their investment. However, given the incremental advancement of science, realizing and accounting for the value of publicly funded research would be extremely difficult. In addition, attaching profit-seeking motives to public entities is unlikely to resolve, and could deepen, the inequalities surrounding access to innovations in health and medicine.

It might be better if finance had no role in health care and science. More specifically, alleviating the effects of infectious diseases will likely need to entail a broader decommodification of health care and science. To make this happen, both would need to be seen as public goods. While many countries throughout the world have publicly funded health care systems, many do not, and some of the existing systems are woefully underfunded.[29] However, places with public health care consistently have populations with better health outcomes, and they are able to provide health care at a lower cost per person.[30] When dealing with infectious diseases, public health care has a number of potential advantages. First, everyone has access to vaccines and treatment, making infectious disease outbreaks less likely to occur and easier to contain. Second, well-funded public health care systems often have more centralized and more robust data on disease occurrence and its impacts on human populations. Public health care systems are thus better able to track where infectious diseases are occurring and spreading. In addition, they can better track, and in turn treat, any long-term effects of infectious diseases.

Turning health and medical science into a public good would entail making the funding and rewards of innovation a more collective enterprise. No patents would be placed on pharmaceuticals or other medical technologies. In addition, such things would need to be produced with minimal to no profit. Proponents of the current system would argue that there would be no incentive for innovation, but most scientists and inventors are driven by the potential to create something new and possibly help society, not by the slim chance of developing a drug or technological advancement that will make them wealthy.[31] Some have argued the opposite: that patents may inhibit innovation, placing undue costs on scientific advancement.[32] To make up for the potential loss of private sector investment, governments would need to increase the amount of public funding available for scientific—in this case medical—advancements. However, without having to pay to use or build on existing patented technologies, the costs of research endeavors would likely go down.

Making health care and science public goods would likely have similar ramifications in each of our cases. While the United States, Brazil, and Uganda have radically different scientific research and health care systems, the proposed actions would benefit the people in all three places. In the United States, creating a robust public health care system would likely make access to treatment easier for those afflicted by Lyme disease and posttreatment Lyme disease. In Brazil, the decommodification of science would lower the price of both innovations intended to reduce populations of mosquitoes and treatment options for mosquito-borne disease. And in Uganda, decommodifying health care and science would make access to medical professionals and treatment for malaria possible for greater segments of the population.

A public health care system in the United States could help those affected by Lyme disease in two primary ways. First, it is plausible that the financial incentives to deny care and treatment that are embedded in the for-profit health care system would diminish. While evidence-based medicine would still likely guide health care, people would not be denied treatment or be stuck footing the bill when they came down with undiagnosed Lyme disease or when the disease had lingering effects on their health. Second, a public health care system with centralized medical records could help future patients suffering from posttreatment Lyme disease. While some research centers are currently studying posttreatment Lyme disease, many people afflicted by it are left outside the conventional medical system. If patients could more easily get care and doctors could document and report patients' symptoms without reprisal, a more robust set of medical data would be compiled that could be used to better diagnose and treat posttreatment Lyme disease.

Decommoditized science could help those impacted by dengue and Zika in Brazil by making scientific innovations more affordable. No treatments currently exist for dengue or Zika, and diseases that largely affect only populations in the global south often receive scant attention and funding from the pharmaceutical industry. However, since the early twentieth century, Brazil has often been at the forefront of developing ways to prevent and treat infectious diseases. Frequently constrained by small budgets, such efforts have waxed and waned at different points in history. To aid these efforts in Brazil, patents could be removed from pharmaceuticals and other potential infectious disease interventions. This reform could allow Brazil to produce and advance its own vaccines, treatments, and other preventive measures—a shot, a pill, or a genetically modified mosquito. This innovation would not only aid populations that frequently suffer from *Ae. aegypti*-borne diseases but also allow for Brazilian scientists' ideas to be better incorporated in the global production of knowledge.

Decommodifying health care and science could help those continually haunted by the specter of malaria in Uganda as well. After independence, Uganda experienced noticeable gains in the overall health of its population as it made dedicated efforts to build hospitals and as the WHO aided in the provision of essential medicines. However, such improvements were derailed during the country's dictatorship and with its subsequent implementation of structural adjustment programs. Decommodifying health care in Uganda could make treatment and medicines less expensive. Decommodifying science could have impacts similar to those possible in Brazil, allowing for a revitalization of domestic pharmaceutical production. The decommodification of both health care and science would also potentially enhance the ready availability of conventional doctors and medicines and in turn decrease the tendency among the poor to "treat from home."

## THE BANK OF THE PLANET

To the unwitting observer, Kim Stanley Robinson's fictional discussion of Wall Street financiers as parasitic killers could seem extreme.[33] So too might the suggestion made in the *Financial Times* that financiers be wiped out. However, financiers and their actions have been defined and labeled as parasitic by Friedrich Hayek and Vladimir Lenin alike. Hayek asserted that financiers often drove the "parasitic circulation" of money from which they gleaned profits outside the realm of production, and Lenin noted in his treatise on imperialism that the export of capital by financiers "sets the seal of parasitism on the whole country that lives by exploiting the labour of several overseas countries and colonies." John Maynard Keynes, whose ideas essentially guided postwar capitalism, saw financiers as essentially functionless investors and coined the phrase "euthanasia of the rentier class."[34] While finance and financiers have thus long been portrayed as partaking in parasitic activities, neither Hayek, Lenin, nor Keynes equated the denizens of Wall Street with actual biological parasites. In each of their conceptualizations, the host on which the financial parasites lived was capitalist society.

Finance has been seen as parasitic largely because it extracts value from society with little return. For some, this is value taken from businesses involved in processes of production. For others, the value is taken from the workers who make the goods and provide the services society needs. According to economic geographer Brett Christophers, the extractive properties of finance negatively affect capitalist society in two ways.[35] First, financial rentierism can lead to economic stagnation and low levels of innovation. As firms focus on paying off debt or increasing shareholder value, the means of competition and profit change from who can create a better and more innovative product to whose stock has the highest return on Wall Street. In other words, the economic incentive to actively innovate no longer exists. Second, financialization can lead to worker exploitation. With shareholder value being the paramount form of competition, both the drive to increase production or make a better product and the ability of workers to use their skills as leverage diminishes. In turn, value increasingly comes at the expense of workers and their wages instead of from the goods or services they produce.

Finance capital can be seen as parasitic in another way, as well. It seeks to extract value from public goods. One way this occurs is by attempting to turn public infrastructure into a profitable financial instrument. For example, profit can be made from the interest rates on infrastructure bonds or through the privatization and sale of shares in a previously public water, sewage, or health care system. Another way is to turn nature into a potentially profitable investment option. For example, profit can be obtained from owning a wetlands mitigation bank or speculatively investing in oil, gold, or another natural resource through a commodities exchange. Through such activities, the extractive properties of finance negatively affect capitalist society by altering and impairing landscapes. In turn, the parasitic behaviors of finance create the conditions for some pathogens to thrive. Finance can thus not only be parasitic but also be pathogenic.

In his fictional commentary on finance and financiers, Kim Stanley Robinson explores the meaning of Keynes's idea "the euthanasia of the rentier class."[36] As Robinson points out, the term *euthanasia* usually refers to helping a person die in order to end their suffering. But, as Robinson further notes, the only thing that members of the financial class potentially suffer from is the guilt, anxiety, depression, and shame they may feel from knowing the harm they have inflicted on people and the planet. The democratization of finance and its reorientation to account for pathogenic ecologies would temper the profits accrued by the financial class and might alleviate some of its mental suffering, but it would not likely result in its death. Furthermore, decommodifying health care and science and turning both into public goods would put a damper on two highly profitable avenues of accruing shareholder value and in turn wealth, but many others would still exist.

To fully end the parasitic and pathogenic tendencies of finance would perhaps require something different. Anarchist Pierre-Joseph Proudhon once suggested that the parasitic properties of finance could be overcome through the creation of a "Bank of the People."[37] However, in Proudhon's original description of the bank, he failed to fully account for nature. That said, a bank could be created whose mission would be to regenerate the planet's ecology in ways that enhanced both biodiversity and human flourishing. At such a bank, credit worthiness would not be determined by who could best turn a profit but by who could create landscapes and institutions that improved the health and well-being of the Earth and all its occupants. One possible name for such an institution could be the Bank of the Planet.

# Notes

ACKNOWLEDGMENTS

1. See Kaup, "Making of Lyme Disease"; Kaup, "Pathogenic Metabolism"; Kaup, Abel, and Sikirica, "Individualized Environments, Individual Cures."


INTRODUCTION

1. O'Keefe, "Farmer Bill."

2. McDonald's, "100 Circle Farms."

3. By way of contrast, according the United States Department of Agriculture, the average cropland value in the United States in 2018 was $4,130 per acre and in Washington state was $2,920. See United States Department of Agriculture National Agricultural Statistics Service, "Land Values 2018 Summary 08/02/2018."

4. O'Keefe, "Farmer Bill."

5. The Bill & Melinda Gates foundation was originally named William H. Gates Foundation. Its name was changed in 2000.

6. United States Security and Exchange Commission, "Bill & Melinda Gates Foundation Form 13F."

7. Bill & Melinda Gates Foundation, "Foundation Fact Sheet."

8. Bill & Melinda Gates Foundation, "About."

9. Bill & Melinda Gates Foundation, "About Grand Challenges."

10. Bill & Melinda Gates Foundation, "Gates Foundation Commits $258.3 Million for Malaria Research and Development."

11. Bill & Melinda Gates Foundation, "Bill and Melinda Gates Call for New Global Commitment to Chart a Course for Malaria Eradication."

12. Jecker and Atuire, "What's Yours Is Ours."

13. Jones et al., "Global Trends in Emerging Infectious Diseases"; Weiss and McMichael, "Social and Environmental Risk Factors."

14. Johns Hopkins Coronavirus Resource Center, "Mortality Analyses"; Troeger, "Just How Do Deaths Due to COVID-19 Stack Up?"

15. Ali, Connolly, and Keil, *Pandemic Urbanism.*

16. Frierson, "The Yellow Fever Vaccine"; Aminov, "Brief History of the Antibiotic Era."

17. Stapleton, "Lessons of History?"

18. Santosa et al., "Development and Experience of Epidemiological Transition Theory."

19. UNAIDS, "Global HIV and AIDS Statistics."

20. Jones et al., "Global Trends in Emerging Infectious Diseases"; Weiss and McMichael, "Social and Environmental Risk Factors."

21. Ilic and Ilic, "Global Patterns of Trends in Cholera Mortality"; World Health Organization, "Summary of Tuberculosis Data."

22. Lowe et al., "Zika Virus Epidemic in Brazil."

23. World Health Organization, "Yellow Fever."

24. World Health Organization, *World Malaria Report 2022.*

25. Patz et al., "Impact of Regional Climate Change on Human Health"; Altizer et al., "Climate Change and Infectious Diseases."

26. Mora et al., "Over Half of Known Human Pathogenic Diseases."

27. Gibb et al., "Zoonotic Host Diversity Increases."

28. Estoque et al., "Spatiotemporal Pattern of Global Forest Change."

29. Ostfeld and Keesing, "Effects of Host Diversity on Infectious Disease"; Schmidt and Ostfeld, "Biodiversity and the Dilution Effect in Disease Ecology."

30. Allan, Felicia, and Ostfeld, "Effect of Forest Fragmentation on Lyme Disease Risk"; Snyder et al., "Zika."

31. Jones et al., "Global Trends in Emerging Infectious Diseases"; Weiss and McMichael, "Social and Environmental Risk Factors."

32. Woolhouse et al., "Global Disease Burden Due to Antibiotic Resistance."

33. Hemingway, Field, and Vontas, "Overview of Insecticide Resistance."

34. Antony and Parija, "Antimalarial Drug Resistance"; Conrad and Rosenthal, "Antimalarial Drug Resistance in Africa."

35. Asidi et al., "Loss of Household Protection from Use of Insecticide-Treated Nets"; Glunt et al., "Empirical and Theoretical Investigation into the Potential Impacts of Insecticide Resistance."

36. Siviter and Muth, "Do Novel Insecticides Pose a Threat to Beneficial Insects?

37. Marx, *Eighteenth Brumaire of Louis Bonaparte*, chapter 1.

38. Tett, "We Should All Be Worried about the 'Financialisation' of Our World."

39. Emba, "Has Our Economy Become Too 'Financialized'?"

40. Denning, "Why Financialization Has Run Amok."

41. Krippner, *Capitalizing on Crisis*, 174. See also Arrighi, *Long Twentieth Century*.

42. Epstein, *Financialization and the World Economy*, 3. See also Davis and Kim, "Financialization of the Economy."

43. Krippner, *Capitalizing on Crisis*.

44. Most scholars see the abandonment of the gold standard as having occurred for two reasons. First, the United States became increasingly concerned that offshore US dollar holdings surpassed the value of its gold reserves, threatening both the value of the currency and the legitimacy of the international monetary system. See McMichael, *Development and Social Change*. Second, adopting a floating exchange rate provided the United States with greater control over its monetary supply and policy, adding to its toolkit a way to help offset inflationary pressures at the time. See Krippner, *Capitalizing on Crisis*.

45. McMichael, *Development and Social Change*.

46. Leyshon and Thrift, "Capitalization of Almost Everything"; Langley, *Everyday Life of Global Finance*; Krippner, *Capitalizing on Crisis*; Helleiner, *States and the Reemergence of Global Finance*.

47. Fairbairn, *Fields of Gold*; Leyshon and Thrift, "Capitalization of Almost Everything."

48. Quark, *Global Rivalries*; Talbot, *Grounds for Agreement*.

49. McMichael, "World-Systems Analysis, Globalization, and Incorporated Comparison."

50. Krippner, *Capitalizing on Crisis*.

51. Krippner, *Capitalizing on Crisis*; Helleiner, *States and the Reemergence of Global Finance*.

52. Gotham, "Secondary Circuit of Capital Reconsidered"; Langley, "Securitising Suburbia."

53. Hall, "Geographies of Money and Finance II"; Langley, *Everyday Life of Global Finance*.

54. Fairbairn, *Fields of Gold*.

55. Duca and Walker, "Why Has U.S. Stock Ownership Doubled?"; Gallup Inc., "U.S. Stock Ownership Highest since 2008."

56. International Monetary Fund, "2023 Global Debt Monitor."

57. Credit Suisse Research Institute, "Global Wealth Report 2023."

58. Lapavistas, "The Government Isn't to Blame for the Rise of Wall Street"; Hudson, "We Can't Save the Economy Unless We Fix Our Debt Addiction"; Lardner, "Are We Repeating History?"; Mukunda, "What Both Bernie Sanders and Donald Trump Get Wrong"; Admati, "In Banking, It's All Other People's Money."

59. Lardner, "Are We Repeating History?"

60. Mukunda, "What Both Bernie Sanders and Donald Trump Get Wrong."

61. Robbins, *Political Ecology.*

62. Gandy, "Zoonotic Urbanization"; Gandy, "Zoonotic City"; Ali, Connolly, and Keil, *Pandemic Urbanism*; Wallace et al., *Clear-Cutting Disease Control*; Ferring and Hausermann, "Political Ecology of Landscape Change, Malaria, and Cumulative Vulnerability."

63. King, *States of Disease*; Neely, "Internal Ecologies and the Limits of Local Biologies."

64. Wallace et al., "The Dawn of Structural One Health"; Wallace et al., *Clear-Cutting Disease Control*; Wallace, *Dead Epidemiologists.*

65. Wallace et al., "Dawn of Structural One Health." 70.

66. Aalbers, "Financial Geography III"; Keil, *Suburban Planet*; Fairbairn, *Fields of Gold.*

67. Robinson, *Market in Mind*; Roy, *Capitalizing a Cure.*

68. Block and Hockett, *Democratizing Finance.*

69. Brown, Morello-Frosch, and Zavestoski, *Contested Illnesses.*

70. Lima et al., "Evidence for an Overwintering Population of *Aedes aegypti.*"

71. Bunker, *Underdeveloping the Amazon*; Clark and Foster, "Ecological Imperialism and the Global Metabolic Rift"; Jorgenson, Austin, and Dick, "Ecologically Unequal Exchange and the Resource Consumption / Environmental Degradation Paradox."

72. Krishna, *One Illness Away*; Portes, "Housing Policy, Urban Poverty, and the State."

73. McMichael, "World-Systems Analysis, Globalization, and Incorporated Comparison"; McMichael, "Incorporating Comparison within a World-Historical Perspective."

74. McMichael, "Incorporating Comparison within a World-Historical Perspective," 392.

75. McMichael, "World-Systems Analysis, Globalization, and Incorporated Comparison."

76. Han, Kramer, and Drake, "Global Patterns of Zoonotic Disease in Mammals."

## CHAPTER 1. A NORTHERN INVASION

1. Rozell and Wilcox, *God at the Grass Roots.*

2. Murphy, "Is Romney Using Lyme Disease to Win Swing State Votes?"; Helmuth, "Romney Targets Lyme Disease Conspiracy Theorists"; Specter, "Mitt Romney versus Lyme Disease and Science."

3. Brown, Kroll-Smith, and Gunter, "Knowledge, Citizens, and Organizations."

4. Eisen and Eisen, "Blacklegged Tick, *Ixodes scapularis*."

5. Bishopp and Trembley, "Distribution and Hosts of Certain North American Ticks."

6. Eisen and Eisen, "Blacklegged Tick, *Ixodes scapularis*."

7. Elbaum-Garfinkle, "Close to Home."

8. Kugeler et al., "Geographic Distribution and Expansion of Human Lyme Disease"; McPherson et al., "Expansion of the Lyme Disease Vector"; United States Environmental Protection Agency, "Climate Change Indicators."

9. Heimberger et al., "Epidemiology of Lyme Disease in Virginia."

10. Eisen, Eisen, and Beard, "County-Scale Distribution of *Ixodes scapularis* and *Ixodes pacificus*."

11. Eisen, Eisen, and Beard.

12. Virginia Department of Health, "Tables of Selected Reportable Diseases in Virginia."

13. Van Zee et al., "Nuclear Markers Reveal Predominantly North to South Gene Flow in *Ixodes scapularis*."

14. Walter et al., "Genomic Insights into the Ancient Spread of Lyme Disease."

15. Spielman, "Emergence of Lyme Disease and Human Babesiosis in a Changing Environment."

16. Berger et al., "Adverse Moisture Events Predict Seasonal Abundance of Lyme Disease Vector Ticks"; Rodgers, Zolnik, and Mather, "Duration of Exposure to Suboptimal Atmospheric Moisture."

17. Eisen and Eisen, "Blacklegged Tick, *Ixodes scapularis*."

18. Ostfeld, *Lyme Disease*.

19. Ostfeld et al., "Tick-Borne Disease Risk in a Forest Food Web"; Vail and Smith, "Vertical Movement and Posture of Blacklegged Tick (Acari: Ixodidae) Nymphs."

20. Rollend, Fish, and Childs, "Transovarial Transmission of Borrelia Spirochetes by *Ixodes scapularis*.

21. Ostfeld, *Lyme Disease*.

22. Cronon, *Changes in the Land*; Greeley, "Relation of Geography to Timber Supply."

23. Dickson, "Wildlife of Southern Forests"; Gray, *Wildlife and People*; Cronon, *Changes in the Land*.

24. Barbour, *Lyme Disease*.

25. Dickson, "Wildlife of Southern Forests"; Gray, *Wildlife and People*.

26. Wood and Lafferty, "Biodiversity and Disease"; LoGiudice et al., "Ecology of Infectious Disease"; Schmidt and Ostfeld, "Biodiversity and the Dilution Effect in Disease Ecology."

27. Dickson, "Wildlife of Southern Forests Habitat and Management"; Cronon, *Changes in the Land*.

28. Gleim et al., "The Phenology of Ticks and the Effects of Long-Term Prescribed Burning."

29. Greeley, "The Relation of Geography to Timber Supply."

30. Cronon, *Changes in the Land*.

31. Gray, *Wildlife and People*.

32. Kimball and Johnson, "The Richness of American Wildlife."

33. Gray, *Wildlife and People*.

34. Dickson, "Wildlife of Southern Forests Habitat and Management"; McCabe and McCabe, "Of Slings and Arrows"; Seton, *Lives of Game Animals*.

35. Barbour, *Lyme Disease*.

36. Wood and Lafferty, "Biodiversity and Disease"; LoGiudice et al., "Ecology of Infectious Disease"; Schmidt and Ostfeld, "Biodiversity and the Dilution Effect in Disease Ecology."

37. Both Dickson and Cronon discuss how hunting during the colonial era in the United States led to the near extinction of a number of animals. See Dickson, "Wildlife of Southern Forests," and Cronon, *Changes in the Land*.

38. Keil, "Extended Urbanization, 'Disjunct Fragments,' and Global Suburbanisms."

39. Schmidt, "Sprawl."

40. Quinn, *American Bonds*.

41. Quinn.

42. Calavita, Pontell, and Tillman, *Big Money Crime*.

43. Hoffmann, *Politics and Banking*.

44. Radford, *Modern Housing for America*.

45. Krippner, *Capitalizing on Crisis*.

46. Mason, *From Buildings and Loans to Bail-Outs*.

47. Krippner, *Capitalizing on Crisis*.

48. Robbins, *Lawn People*; Jackson, *Crabgrass Frontier*.

49. Davison, "Suburban Idea and Its Enemies"; Melosi, *Sanitary City*.

50. Krippner, *Capitalizing on Crisis*.

51. Krippner.

52. Nocera, *Piece of the Action*; Krippner, *Capitalizing on Crisis*.

53. Meyerson, "Changing Structure of Housing Finance in the United States"; Florida, "The Political Economy of Financial Deregulation."

54. United States Department of Housing and Urban Development, "Survey of Mortgage Lending Activity."

55. Federal Reserve, "Mortgage Debt Outstanding."

56. Federal Housing Finance Agency, "National Average Contract Mortgage Rate History."

57. Federal Housing Finance Agency.

58. United States Census Bureau, "Housing Vacancies and Homeownership."

59. Segal and Sullivan, "Trends in Homeownership."

60. United States Census Bureau, "Median and Average Square Feet of Floor Area in New Single-Family Houses Sold."

61. Heimlich and Anderson, "Development at the Urban Fringe and Beyond."

62. Bruegmann, *Sprawl.*

63. Ostfeld, *Lyme Disease.*

64. Ostfeld.

65. LoGiudice et al., "Ecology of Infectious Disease"; Allan, Keesing, and Ostfeld, "Effect of Forest Fragmentation on Lyme Disease Risk."

66. Kulikoff, *Tobacco and Slaves.*

67. Kulikoff, *Tobacco and Slaves*; Sweig, "Importation of African Slaves to the Potomac River."

68. Greeley, "Relation of Geography to Timber Supply."

69. Thompson and Francl-Powers, "Brief History of Terrestrial Game Species Management."

70. Virginia Department of Game and Inland Fisheries, "Virginia Deer Management Plan."

71. Virginia Department of Game and Inland Fisheries.

72. Richards, "Debates over the Retrocession of the District of Columbia."

73. Friedman, *Covert Capital.*

74. Fairfax County GIS & Mapping, "Fairfax County Historical Imagery Viewer."

75. Friedman, *Covert Capital.*

76. Netherton et al., *Fairfax County, Virginia.*

77. Friedman, *Covert Capital.*

78. Friedman.

79. Minnesota Population Center, "National Historical Geographic Information System."

80. Netherton et al., *Fairfax County, Virginia.*

81. Friedman, *Covert Capital.*

82. Henig, "Privatization in the United States"; Baber, "Privatizing Public Management."

83. Friedman, *Covert Capital.*

84. Friedman.

85. Minnesota Population Center, "National Historical Geographic Information System."

86. Clapson, "Suburban Paradox?"; Daniels, *When City and Country Collide.*

87. Virginia Department of Game and Inland Fisheries, "Virginia Deer Management Plan."

88. Virginia Department of Game and Inland Fisheries.

89. Heimberger et al., "Epidemiology of Lyme Disease in Virginia."

90. Virginia Department of Health, "Tables of Selected Reportable Diseases in Virginia."

91. Virginia Department of Health.

## CHAPTER 2. THE WORST ANIMAL IN THE WORLD

1. Sokol, "Worst Animal in the World."

2. Dorit, "Zika Goes Viral."

3. Powell, Gloria-Soria, and Kotsakiozi, "Recent History of *Aedes aegypti.*"

4. Sokol, "Worst Animal in the World."

5. Beaubien, "This Mosquito Likes Us Too Much for Our Own Good."

6. Powell, Gloria-Soria, and Kotsakiozi, "Recent History of *Aedes aegypti.*"

7. McNeill, *Plagues and Peoples.*

8. Centers for Disease Control and Prevention, "Global Health—Newsroom—Yellow Fever."

9. World Health Organization, "Dengue and Severe Dengue."

10. Centers for Disease Control and Prevention, "Dengue Clinical Case Management (DCCM)."

11. Impieri Souza et al., "Geography of Microcephaly in the Zika Era"; Worth and Mizner, "Zika Uncontained."

12. Worth and Mizner, "Zika Uncontained."

13. Impieri Souza et al., "Geography of Microcephaly in the Zika Era."

14. Falcão Sobral and da Penha Sobral, "Casos de dengue e coleta de lixo urbano"; Kikuti et al., "Spatial Distribution of Dengue in a Brazilian Urban Slum Setting."

15. Ministério de Saúde, "Zika Virus."

16. Catarina Andrioli, Assunta Busato, and Antonio Lutinski, "Spatial and Temporal Distribution of Dengue in Brazil"; Teixeira et al., "Dengue."

17. Ioris, *Agribusiness and the Neoliberal Food System in Brazil*; Yamada, "Cerrado of Brazil."

18. Brown et al., "Human Impacts Have Shaped Historical and Recent Evolution in *Aedes aegypti.*"

19. Powell, Gloria-Soria, and Kotsakiozi, "Recent History of *Aedes aegypti.*"

20. Brady et al., "Global Temperature Constraints on *Aedes aegypti* and *Ae. albopictus.*"

21. Yang et al., "Assessing the Effects of Temperature on the Population of *Aedes aegypti.*"

22. Barredo and DeGennaro, "Not Just from Blood."

23. McBride et al., "Evolution of Mosquito Preference for Humans Linked to an Odorant Receptor."

24. Dorit, "Zika Goes Viral."

25. Powell, Gloria-Soria, and Kotsakiozi, "Recent History of *Aedes aegypti*."

26. Barcia, *Yellow Demon of Fever.*

27. Powell, Gloria-Soria, and Kotsakiozi, "Recent History of *Aedes aegypti*."

28. Espinosa, "The Question of Racial Immunity to Yellow Fever in History and Historiography."

29. Chippaux and Chippaux, "Yellow Fever in Africa and the Americas."

30. Souza Salles et al., "History, Epidemiology, and Diagnostics of Dengue in the American and Brazilian Contexts."

31. Marques et al., "Atlantic Forest"; Brathwaite Dick et al., "History of Dengue Outbreaks in the Americas."

32. Solórzano, Brasil, and Oliveira, "Atlantic Forest Ecological History."

33. Solórzano, Brasil, and Oliveira, "Atlantic Forest Ecological History"; Oliveira, "When the Shifting Agriculture Is Gone."

34. McNeill, "Ecology, Epidemics, and Empires."

35. Metcalf, *Go-Betweens and the Colonization of Brazil.*

36. Metcalf.

37. Solórzano, Brasil, and Oliveira, "Atlantic Forest Ecological History."

38. McNeill, "Agriculture, Forests, and Ecological History."

39. Solórzano, Brasil, and Oliveira, "Atlantic Forest Ecological History."

40. McNeill, *Mosquito Empires.*

41. McNeill, "Ecology, Epidemics and Empires."

42. Metcalf, *Go-Betweens and the Colonization of Brazil.*

43. Metcalf.

44. Metcalf.

45. For a detailed account of the extensive land use change that occurred along the Brazilian coast, see McNeill, "Agriculture, Forests, and Ecological History."

46. Cooper, "Brazil's Long Fight against Epidemic Disease."

47. Skidmore, *Brazil*; Stepan, *Beginnings of Brazilian Science.*

48. Skidmore, *Brazil.*

49. Castro-Santos, "Power, Ideology, and Public Health in Brazil."

50. Castro-Santos.

51. Hochman, *Sanitation of Brazil.*

52. Williams, "Nationalism and Public Health."

53. Williams.

54. Hochman, "From Autonomy to Partial Alignment."

55. Löwy, "Leaking Containers."

56. Franco, *História da Febre-Amarela no Brasil.*

57. Hochman, "From Autonomy to Partial Alignment."

58. Ioris, *Transforming Brazil.*

59. Ioris, *Transforming Brazil.*

60. Ioris, *Transforming Brazil.*

61. Baer, *Brazilian Economy.*

62. Baer.

63. Geddes, "Building 'State' Autonomy in Brazil."

64. Sikkink, *Ideas and Institutions.*

65. Baer and Villela, "Changing Nature of Development Banking in Brazil."

66. Fuentes Sarmiento, *Spatial Distribution of Industrial Production in Brazil.*

67. Instituto Brasileiro de Geografia e Estatística, "Table 1288."

68. Skidmore, *Brazil.*

69. Nuijten, Koster, and de Vries, "Regimes of Spatial Ordering in Brazil."

70. Ramsdell, "National Housing Policy and the Favela in Brazil."

71. Garmany, "Situating Fortaleza"; Nuijten, Koster, and de Vries, "Regimes of Spatial Ordering in Brazil."

72. Ramsdell, "National Housing Policy and the Favela in Brazil."

73. Koster and Nuijten, "From Preamble to Post-Project Frustrations."

74. Presidência da República Casa Civil Subchefia para Assuntos Jurídicos, "Lei No. 4.380, de 21 de Agosto de 1964."

75. Martins, "Justiça autoriza empresa a comercializar *Aedes aegypti* modificado"; Portes, "Housing Policy, Urban Poverty, and the State."

76. Baer and Beckerman, "Trouble with Index-Linking."

77. Souza and Zetter, "Urban Land Tenure in Brazil"; Portes, "Housing Policy, Urban Poverty, and the State."

78. Valladares and Figueiredo, "Housing in Brazil."

79. Valença, "Inevitable Crisis of the Brazilian Housing Finance System."

80. Melo, "State Retreat, Governance, and Metropolitan Restructuring in Brazil."

81. Portes, "Housing Policy, Urban Poverty, and the State."

82. Valladares and Figueiredo, "Housing in Brazil."

83. Perlman, *Favela*; Ramsdell, "National Housing Policy and the Favela in Brazil."

84. Valladares and Figueiredo, "Housing in Brazil."

85. Melo, "State Retreat, Governance, and Metropolitan Restructuring in Brazil."

86. Frieden, *Banking on the World.*

87. Frieden, *Banking on the World.*

88. Baer, *Brazilian Economy.*

89. Melo, "State Retreat, Governance, and Metropolitan Restructuring in Brazil."

90. After the BNH went bankrupt in 1986, its debts were transferred to the Caixa Econômica Federal. See Valença, "Inevitable Crisis of the Brazilian Housing Finance System."

91. Melo, "State Retreat, Governance, and Metropolitan Restructuring in Brazil."

92. Valença, "Inevitable Crisis of the Brazilian Housing Finance System."

93. Melo, "State Retreat, Governance, and Metropolitan Restructuring in Brazil."

94. Melo.

95. O'Hare and Barke, "Favelas of Rio de Janeiro."

96. Lloyd-Sherlock, "Recent Appearance of Favelas in São Paulo City."

97. Nuijten, Koster, and Vries, "Regimes of Spatial Ordering in Brazil."

98. Melo, "State Retreat, Governance, and Metropolitan Restructuring in Brazil*."

99. Löwy, "Leaking Containers."

100. Löwy.

101. Guerra Nunes et al., "30 Years of Fatal Dengue Cases in Brazil."

102. Bosco Siqueira et al., "Dengue and Dengue Hemorrhagic Fever."

103. Catarina Andrioli, Assunta Busato, and Lutinski, "Spatial and Temporal Distribution of Dengue in Brazil."

104. Lowe et al., "Zika Virus Epidemic in Brazil."

105. Zinszer et al., "Reconstruction of Zika Virus Introduction in Brazil."

106. Lowe et al., "Zika Virus Epidemic in Brazil."

107. Impieri Souza et al., "Geography of Microcephaly in the Zika Era."

108. Skidmore, *Brazil*.

109. Klink and Moreira, "Past and Current Human Occupation."

110. Frank, "Elite Families and Oligarchic Politics on the Brazilian Frontier."

111. Klink and Moreira, "Past and Current Human Occupation."

112. Mittermeier et al., *Hotspots Revisited*.

113. Prager and Milhorance, "Cerrado"; Lima, "Public Health and Social Ideas in Modern Brazil."

114. Klink, Moreira, and Solbrig, "Ecological Impact of Agricultural Development in the Brazilian Cerrados."

115. Ioris, *Agribusiness and the Neoliberal Food System in Brazil*.

116. Schmink and Wood, *Frontier Expansion in Amazonia*.

117. Ioris, *Agribusiness and the Neoliberal Food System in Brazil*.

118. Helfand, "Political Economy of Agricultural Policy in Brazil."

119. Ioris, *Agribusiness and the Neoliberal Food System in Brazil*.

120. Ozorio de Almeida, *Colonization of the Amazon*; Ioris, *Agribusiness and the Neoliberal Food System in Brazil*.

121. Moran, "Deforestation and Land Use in the Brazilian Amazon."

122. Helfand, "Political Economy of Agricultural Policy in Brazil"; Ioris, *Agribusiness and the Neoliberal Food System in Brazil*.

123. Helfand, "Political Economy of Agricultural Policy in Brazil"; Helfand and Castro de Rezende, "Impact of Sector-Specific and Economy-Wide Policy Reforms."

124. Wilkinson, Reydon, and Di Sabbato, "Concentration and Foreign Ownership of Land in Brazil."

125. Requiao, "Brazilian Agricultural Export Promotion Experience to Advance Agricultural Trade."

126. Helfand and Castro de Rezende, "Impact of Sector-Specific and Economy-Wide Policy Reforms."

127. Ioris, *Agribusiness and the Neoliberal Food System in Brazil.*

128. Fairbairn, "'Like Gold with Yield.'"

129. United States Department of Agriculture Foreign Agricultural Services, "World Agricultural Production"; United States Department of Agriculture Foreign Agricultural Services, "Commodity Intelligence Report."

130. Instituto Nacional de Pesquisas Espaciais (INPE) Observação da Terra, "Monitoramento do Desmatamento da Floresta Amazônica Brasileira por Satélite."

131. Strassburg et al., "Moment of Truth for the Cerrado Hotspot."

132. Instituto Brasileiro de Geografia e Estatística, "Censo Agropecuário."

133. Instituto Brasileiro de Geografia e Estatística, "Censo Demográfico."

134. Ioris, *Agribusiness and the Neoliberal Food System in Brazil.*

135. Instituto Brasileiro de Geografia e Estatística, "Censo Demográfico."

136. Sawyer et al., "Ecosystem Profile: Cerrado Biodiversity Hotspot"; Chaplin-Kramer et al., "Spatial Patterns of Agricultural Expansion Determine Impacts on Biodiversity and Carbon Storage."

137. Alencar et al., "Distribution of Haemagogus and Sabethes Species in Relation to Forest Cover and Climatic Factors."

138. Sistema Nacional de Informações sobre Saneamento (SNIS), "Série Histórica."

139. Ministério de Saúde, "Zika Virus."

140. SNIS, "Série Histórica."

141. Sokol, "Worst Animal in the World."

CHAPTER 3. ALL THAT REMAINS IS MAN
AND MOSQUITO

1. Watira, "Bududa District Local Government Five-Year District Development Plan."

2. Neafsey et al., "Highly Evolvable Malaria Vectors."

3. Rose et al., "Climate and Urbanization Drive Mosquito Preference for Humans."

4. Benton et al., "Influence of Evolutionary History on Human Health and Disease."

5. World Health Organization, *World Malaria Report 2019*.

6. Kwashirai, "Environmental Change, Control, and Management in Africa."

7. Kwashirai.

8. Kwashirai.

9. Norris, "Mosquito-Borne Diseases as a Consequence of Land Use Change"; Barros and Honório, "Deforestation and Malaria on the Amazon Frontier."

10. Laporta et al., "Biodiversity Can Help Prevent Malaria Outbreaks in Tropical Forests."

11. Masefield, "Agricultural Change in Uganda."

12. Iyamulemye, "History of Coffee in Uganda."

13. De Haas, "Failure of Cotton Imperialism in Africa."

14. Overseas Development Institute, "Annual Report, 1964."

15. Overseas Development Institute.

16. Nyakaana, "Agricultural Development Planning and Policy in Uganda."

17. Nyakaana.

18. Overseas Development Institute, "Annual Report, 1964."

19. Shinyekwa et al., "Evolution of Industry in Uganda"; Nyakaana, "Agricultural Development Planning and Policy in Uganda."

20. Shinyekwa et al., "Evolution of Industry in Uganda."

21. Kaberuka, "Uganda's 1946 Colonial Development Plan."

22. Windel, *Cooperative Rule*.

23. Young, Sherman, and Rose, *Cooperatives and Development*.

24. Young, Sherman, and Rose.

25. De Haas and Papaioannou, "Resource Endowments and Agricultural Commercialization in Colonial Africa."

26. Nyakaana, "Agricultural Development Planning and Policy in Uganda"; Windel, *Cooperative Rule*; de Haas, "Measuring Rural Welfare in Colonial Africa"; Iyamulemye, "History of Coffee in Uganda."

27. De Haas, "Measuring Rural Welfare in Colonial Africa"; Young, Sherman, and Rose, *Cooperatives and Development*.

28. Uganda Protectorate, *Annual Medical and Sanitary Report for the Year 1912* (1.86); Uganda Protectorate, *Annual Report of the Medical Department for the Year Ended 31st December, 1939* (20); Uganda Protectorate, *Annual Report of the Ministry of Health* (30).

29. Talbot, *Grounds for Agreement*.

30. Gwyer, "Three International Commodity Agreements."

31. Several farmers whom Kelly interviewed spoke of how local cooperatives worked in collaboration with the Ugandan Department of Agriculture to educate farmers on how to grow higher-quality coffee and create local cooperative

pulping stations in order to secure greater value for their produce in the global marketplace.

32. Mamdani, *Politics and Class Formation in Uganda.*

33. Young, Sherman, and Rose, *Cooperatives and Development.*

34. Wiegratz, *Neoliberal Moral Economy*; Oloya, "Marketing Boards and Post-War Economic Development Policy in Uganda."

35. Daviron and Ponte, *Coffee Paradox*; Wiegratz, *Neoliberal Moral Economy.*

36. Talisuna et al., "Past, Present, and Future Use of Epidemiological Intelligence"; Sejjaaka, "Political and Economic History of Uganda, 1962–2002."

37. Talisuna et al., "The Past, Present, and Future Use of Epidemiological Intelligence."

38. Masiga and Ruhweza, "Commodity Revenue Management," 9–10; Bunker, *Peasants against the State.*

39. Young, Sherman, and Rose, *Cooperatives and Development.*

40. Asiimwe, "From Monopoly Marketing to Coffee Magendo."

41. Nyakaana, "Agricultural Development Planning and Policy in Uganda."

42. Masiga and Ruhweza, "Commodity Revenue Management"; World Bank, *World Development Report 1983.*

43. Bunker, *Peasants against the State*; Holmgren et al., "Aid and Reform in Uganda."

44.
World Bank, "External Debt Stocks, Total (DOD, Current US$)—Uganda."

45. World Bank, *World Development Report 1983.*

46. World Bank, *World Development Report 1983.*

47. World Bank, *World Development Report 1983*; World Bank, "Agricultural Sector Adjustment."

48. World Bank, "Agricultural Sector Adjustment"; Adenle, Wedig, and Azadi, "Sustainable Agriculture and Food Security in Africa."

49. Baffes, "The Cotton Sector of Uganda."

50. Ponte, "Coffee Markets in East Africa"; Adenle, Wedig, and Azadi, "Sustainable Agriculture and Food Security in Africa."

51. Talbot, *Grounds for Agreement.* Talbot shows how the United States, as the largest consumer of coffee, essentially held veto power over decision making in the International Coffee Council, which oversaw the ICA. By the end of the 1980s, he notes, the premise of the ICA no longer aligned with the United States' embrace of free trade. Opposed to any form of protectionism, the United States thus opposed renewing the ICA.

52. Talbot, "Information, Finance, and the New International Inequality."

53. Talbot. "Information, Finance, and the New International Inequality," 231.

54. Masiga and Ruhweza, "Commodity Revenue Management"; Baffes, *Restructuring Uganda's Coffee Industry.*

55. Masiga and Ruhweza, "Commodity Revenue Management."

56. Uganda Coffee Development Authority, "Coffee Exports from 1964–2021."

57. Uganda Coffee Development Authority, "Uganda Coffee Development Authority Annual Report 2019–2020."

58. Food and Agriculture Organization of the United Nations, "Statistical Yearbook 2021."

59. World Health Organization, *World Malaria Report 2021.*

60. World Health Organization, *World Malaria Report 2022.*

61. World Health Organization, *World Malaria Report 2021.*

62. Bunker, *Peasants against the State*; Mamdani, *Politics and Class Formation in Uganda.*

63. Cavanagh and Himmelfarb, "'Much in Blood and Money'"; Uganda Coffee Development Authority, "Uganda Coffee Development Authority Annual Report 2018–2019" (at least 80%).

64. Laman, Khamati, and Milimo, "Mount Elgon Integrated Conservation and Development Project (MEICDP)."

65. Mugagga, Kakembo, and Buyinza, "A Characterisation of the Physical Properties of Soil."

66. Cavanagh and Himmelfarb, "'Much in Blood and Money.'"

67. Cavanagh and Himmelfarb.

68. Turyahabwe and Banana, "An Overview of History and Development of Forest Policy and Legislation in Uganda."

69. Cavanagh and Himmelfarb, "'Much in Blood and Money.'"

70. Bunker, *Peasants against the State.*

71. Wedig, *Cooperatives, the State, and Corporate Power*; Wiegratz, *Neoliberal Moral Economy.*

72. Wedig, *Cooperatives, the State, and Corporate Power.*

73. Uganda Coffee Development Authority, "Annual Report 2018–2019."

74. Charles, "Consolidation in the Coffee Industry Is Only Just Getting Started."

75. International Trade Center, "Trade Map—List of Exporters"; Intercontinental Exchange, "Historical Monthly Volumes—Futures."

## CHAPTER 4. THE TREATMENT OF FRINGE BENEFITS

1. Wooten, "'The Most Glorious Story of Failure in the Business.'"

2. Grann, "Stalking Dr. Steere."

3. Aronowitz, "Lyme Disease."

4. Gottschalk, *Shadow Welfare State.*

5. Centers for Disease Control and Prevention, "How Many People Get Lyme Disease?"

6. Centers for Disease Control and Prevention, "Signs and Symptoms of Lyme Disease."

7. Centers for Disease Control and Prevention, "Signs and Symptoms of Lyme Disease."

8. Grann, "Stalking Dr. Steere."

9. Aronowitz, "Lyme Disease."

10. Elbaum-Garfinkle, "Close to Home."

11. Aronowitz, "Lyme Disease."

12. Aronowitz.

13. Steere et al., "Erythema Chronicum Migrans and Lyme Arthritis," 696.

14. Aronowitz, "Lyme Disease." 91.

15. Hacker, "The Rise of Reform," 12.

16. Hacker, 13.

17. Gottschalk, *Shadow Welfare State*, 53–54.

18. Hacker, "The Rise of Reform," 13.

19. Gottschalk, *The Shadow Welfare State*, 57–62.

20. Scott et al., *Institutional Change and Healthcare Organizations*, 41–42.

21. Hacker, "The Rise of Reform," 15.

22. Kaiser Family Foundation, "Employer Health Benefits Survey."

23. Wright and Perry, "Medical Sociology and Health Services Research."

24. Mechanic, "Managed Care Backlash."

25. Hacker, "The Rise of Reform," 15.

26. Reischauer, "Statement of Robert D. Reischauer."

27. Bolen and Hall, "Managed Care and Evidence-Based Practice."

28. Sackett et al., "Evidence Based Medicine."

29. Timmermans and Berg, *Gold Standard*, 14–15.

30. Denny, "Evidence-Based Medicine and Medical Authority"; Brook, "Practice Guidelines and Practicing Medicine."

31. Timmermans and Berg, *Gold Standard*, 16.

32. Sackett et al., "Evidence Based Medicine"; Timmermans and Berg, *Gold Standard*, 16–17.

33. Field and Lohr, *Clinical Practice Guidelines*, 1, 5. In 1999, the Agency for Health Care Policy and Research (AHCPR) became the Agency for Healthcare Research and Quality (AHRQ).

34. Field and Lohr. 3.

35. Agency for Healthcare Research and Quality, "Clinical Guidelines and Recommendations"; Agency for Healthcare Research and Quality, "Evidence-Based Practice Centers"; Dumes, *Divided Bodies*.

36. Siu, Bibbins-Domingo, and Grossman, "Evidence-Based Clinical Prevention."

37. Feinstein and Horwitz, "Problems in the 'Evidence' of 'Evidence-Based Medicine.'"

38. Willis and White, "Evidence Based Medicine, the Medical Profession, and Health Policy"; Feinstein and Horwitz, "Problems in the 'Evidence' of 'Evidence-Based Medicine.'"

39. Timmermans and Berg, *Gold Standard*, 16.

40. Ioannidis, "Evidence-Based Medicine Has Been Hijacked," 82.

41. Sears et al., "Charting the Path Forward."

42. Lyme and Tick-Borne Diseases Research Center, "Diagnosis"; Liang et al., "Rapid Clearance of *Borrelia burgdorferi* from the Blood Circulation."

43. Berger et al., "Cultivation of *Borrelia burgdorferi* from Erythema Migrans Lesions and Perilesional Skin"; Mead, Peterson, and Hinckley, "Updated CDC Recommendation for Serologic Diagnosis of Lyme Disease."

44. Dumes, *Divided Bodies*.

45. Moore et al., "Current Guidelines, Common Clinical Pitfalls, and Future Directions."

46. Wormser et al., "The Clinical Assessment, Treatment, and Prevention."

47. Lantos et al., "Clinical Practice Guidelines"; Centers for Disease Control and Prevention, "Lyme Disease Treatment."

48. National Institute of Allergy and Infectious Diseases, "Chronic Lyme Disease."

49. For a more detailed examination of the politics surrounding the IDSA guidelines and the various actors upholding and contesting them, see Dumes, *Divided Bodies*.

50. Lantos et al., "Clinical Practice Guidelines"; Centers for Disease Control and Prevention, "Preventing Tick Bites on People."

51. Centers for Disease Control and Prevention, "Preventing Ticks in the Yard."

52. Lantos et al., "Clinical Practice Guidelines," 3.

53. Kaup, Abel, and Sikirica, "Individualized Environments, Individual Cures."

54. Agency for Healthcare Research and Quality, "Medical Expenditure Panel Survey"

55. Wooten, "Most Glorious Story of Failure in the Business"; *Business Week*, "New Try for S-P."

56. *Business Week*, "Studebaker-Packard Bets Its All"; *Wall Street Journal*, "Studebaker Picks New Chief"; *Wall Street Journal*, "Metropolitan Life Sells Its Studebaker Convertible Preferred."

57. Carroll and Arnett, "Private Health Insurance Plans in 1978 and 1979"; Gabel, "Marketwatch: Ten Ways HMOs Have Changed during the 1990s."

58. Data obtained from macrotrends.net.

59. New York Stock Exchange, "United Health Group Inc. (UNH)."

CHAPTER 5. MOSQUITO DERIVATIVES

1. *Oxitec's Commercial Launch in Brazil.*

2. Carvalho et al., "Suppression of a Field Population of Aedes Aegypti in Brazil."

3. Pan American Health Association, "Timeline: Emergence of Zika Virus in the Americas."

4. Hogan Lovells International, LLP, "Agreement for the Acquisition of the Entire Issued and to Be Issued Share Capital of Oxitec Limited."

5. Nasdaq, "Precigen, Inc. Common Stock (PGEN) Charts."

6. Centers for Disease Control and Prevention, "Dengue Clinical Case Management"; Centers for Disease Control and Prevention, "Global Health—Newsroom—Yellow Fever"; Centers for Disease Control and Prevention, "Dengue Clinical Case Management."

7. World Health Organization, "One Year into the Zika Outbreak"; Antoniou et al., "Zika Virus and the Risk of Developing Microcephaly."

8. Williams, "Nationalism and Public Health"; Löwy, "Epidemiology, Immunology, and Yellow Fever."

9. Benchimol, "Yellow Fever Vaccine in Brazil."

10. Löwy, "Leaking Containers."

11. Shearer et al., "Global Yellow Fever Vaccination Coverage from 1970 to 2016."

12. Löwy, "Leaking Containers."

13. Feldman Marzochi, "Dengue in Brazil."

14. Pires-Alves, Assunção Paiva, and Lima, "Na Baixada Fluminense, à Sombra da 'Esfinge do Rio.'"

15. Pinto, Cerbino Neto, and Penna, "Evolution of the Federal Funding Policies for the Public Health Surveillance Component."

16. World Bank, "Loan Agreement 2061."

17. World Bank, "Report No. 18142 Brazil."

18. Elias and Cohn, "Health Reform in Brazil."

19. Macinko and Harris, "Brazil's Family Health Strategy."

20. Elias and Cohn, "Health Reform in Brazil."

21. Frizon Rizzotto and de Sousa Campos, "World Bank and the Brazilian National Health System."

22. Pinto, Neto, and Penna, "Evolution of the Federal Funding Policies for the Public Health Surveillance Component."

23. Braga and Valle, "*Aedes aegypti*"; Araújo et al., "*Aedes aegypti* Control Strategies in Brazil."

24. Braga and Valle, "*Aedes Aegypti.*"

25. Macinko and Lima-Costa, "Horizontal Equity in Health Care Utilization in Brazil."

26. Guerra Nunes et al., "30 Years of Fatal Dengue Cases in Brazil."

27. Do Rosário Costa, "Municipalização do controle da doença foi equívoco que custou caro ao país."

28. Do Rosário Costa.

29. Alphey et al., "Sterile-Insect Methods for Control of Mosquito-Borne Diseases."

30. Klassen, Curtis, and Hendrichs, "History of the Sterile Insect Technique."

31. Thomas et al., "Insect Population Control Using a Dominant, Repressible, Lethal Genetic System."

32. Savage, "Spin-Out Fever"; Isis Innovation, "Starting a Spinout Company," 12.

33. Glaser, "Interview with Luke Alphey, Ph.D."; Thomas et al., "Insect Population Control Using a Dominant, Repressible, Lethal Genetic System."

34. Olson et al., "Genetic Approaches in *Aedes aegypti* for Control of Dengue"; Harris et al., "Field Performance of Engineered Male Mosquitoes."

35. Harris et al., "Successful Suppression of a Field Mosquito Population by Sustained Release of Engineered Male Mosquitoes"; Carvalho et al., "Suppression of a Field Population of *Aedes aegypti* in Brazil by Sustained Release of Transgenic Male Mosquitoes."

36. United States Patent and Trademark Office, "Trademark Electronic Search System."

37. Messina et al., "Current and Future Global Distribution and Population at Risk of Dengue"; Alphey et al., "Sterile-Insect Methods for Control of Mosquito-Borne Diseases."

38. Harris et al., "Field Performance of Engineered Male Mosquitoes"; Carvalho et al., "Suppression of a Field Population of *Aedes aegypti* in Brazil by Sustained Release of Transgenic Male Mosquitoes."

39. Garziera et al., "Effect of Interruption of Over-Flooding Releases of Transgenic Mosquitoes over Wild Population of *Aedes aegypti*"; Servick, "Brazil Will Release Billions of Lab-Grown Mosquitoes to Combat Infectious Disease."

40. So as not to be confused with the Islamic State of Iraq and Syria (ISIS), a terrorist organization according to the U.S. State Department, Oxford renamed Isis Innovation to Oxford University Innovation Limited.

41. Oxford University Innovation, "Oxford Spinout Oxitec Sold to Intrexon Corporation for $160 Million."

42. While Oxford does not allow its academics to be directly employed by any of its spin-outs, academics work on consultancy contracts that specify their duties and the amount of time they spend aiding the venture. Within this arrangement, scientists operate as researchers, shareholders, and sometimes directors of the spin-out. From the founding of Oxitec, Luke Alphey was the chief scientific officer. He initially worked in a consultancy role while maintaining his position

at Oxford. In March 2008, he resigned his Oxford position and went to work full-time at Oxitec. His colleague David Kelly cofounded the spin-out and served as its original director. Paul G. Coleman, also a cofounder, served as the business development manager. In 2006, Kelly and Coleman left Oxitec and founded H2O Impact Ventures, a venture capital firm that seeks to commercialize early-stage technologies. Alphey stayed on at Oxitec and later became its director in 2014. He remained in that role until it was sold to Intrexon the following year.

43. Hogan Lovells International LLP, "Agreement for the Acquisition of the Entire Issued and to Be Issued Share Capital of Oxitec Limited."

44. Isis Innovation, "New Technique to Control Killer Disease Offered Multi Million Dollar Funding"; Oxitec Ltd., "Oxitec Receives US$6.8 Million from Wellcome Trust to Advance Scale-up of Friendly™ Aedes Aegypti Technology."

45. Bill & Melinda Gates Foundation, "Committed Grants."

46. Marshall, "Cartagena Protocol in the Context of Recent Releases of Transgenic and Wolbachia-Infected Mosquitoes."

47. Sica de Campos et al., "Responsible Innovation and Political Accountability."

48. Reis-Castro and Hendrickx, "Winged Promises"; Sica de Campos et al., "Responsible Innovation and Political Accountability."

49. Sica de Campos et al., "Responsible Innovation and Political Accountability."

50. Reis-Castro and Hendrickx, "Winged Promises."

51. Carvalho et al., "Suppression of a Field Population of *Aedes aegypti* in Brazil by Sustained Release of Transgenic Male Mosquitoes."

52. Sica de Campos et al., "Responsible Innovation and Political Accountability."

53. Servick, "Brazil Will Release Billions of Lab-Grown Mosquitoes to Combat Infectious Disease."

54. Sica de Campos et al., "Responsible Innovation and Political Accountability."

55. Sica de Campos et al.

56. Sica de Campos et al.

57. Sica de Campos et al.; Oxitec Ltd., "Oxitec Announces Friendly™ Aedes Project Is Supported by 92.8% of Piracicaba's Population."

58. Ferreira, "Inside the Mosquito Factory That Could Stop Dengue and Zika."

59. Ferreira.

60. Moraes, "Justiça autoriza Oxitec a comercializar mosquito transgênico."

61. Martins, "Justiça autoriza empresa a comercializar Aedes aegypti modificado."

62. Oxitec Ltd., "Oxitec Launches Field Trial in Brazil for Next Generation Addition to Friendly™ Mosquitoes Platform"; Oxitec Ltd., "Oxitec Successfully

Completes First Field Deployment of 2nd Generation Friendly™ Aedes Aegypti Technology."

63. Oxitec Ltd., "Oxitec Transitioning Friendly™ Self-Limiting Mosquitoes to 2nd Generation Technology Platform."

64. Oxitec Ltd., "Oxitec Announces Ground-Breaking Commercial Launch of Its Friendly™ *Aedes aegypti* Solution in Brazil."

65. Marin, "Empresa lança assinatura de 'mosquitos do bem' que combatem dengue."

66. Oxitec Ltd., "Oxitec Announces Ground-Breaking Commercial Launch of Its Friendly™ *Aedes aegypti* Solution in Brazil."

67. Chouin-Carneiro et al., "Zika Virus Transmission by Brazilian *Aedes aegypti* and *Aedes albopictus* Is Virus Dose and Temperature-Dependent."

68. Alto et al., "Larval Competition Differentially Affects Arbovirus Infection in *Aedes* Mosquitoes."

69. Servick, "Brazil Will Release Billions of Lab-Grown Mosquitoes to Combat Infectious Disease?"

70. Nasdaq, "Precigen, Inc. Common Stock (PGEN) Advanced Charting."

71. Intrexon, "Intrexon to Achieve $175M Cash Goal."

## CHAPTER 6. TREATING FROM HOME

1. Maude, Woodrow, and White, "Artemisinin Antimalarials"; McNeil, "New Drug for Malaria Pits U.S. against Africa"; McNeil, "For Intrigue, Malaria Drug Gets the Prize"; Shah, *The Fever*.

2. McNeil, "New Drug for Malaria Pits U.S. Against Africa"; Shah, *The Fever*.

3. McNeil, "New Drug for Malaria Pits U.S. Against Africa"; Shah, *The Fever*.

4. Shah, *The Fever*.

5. World Health Organization, *Good Procurement Practices for Artemisinin-Based Antimalarial Medicines*.

6. World Health Organization.

7. Shah, *The Fever*.

8. Pou et al., "Sontochin as a Guide to the Development of Drugs Against Chloroquine-Resistant Malaria."

9. It was also extremely difficult to determine proper dosing for quinine, and the drug has many adverse side-effects, including tinnitus, deafness, headache, nausea, visual disturbances, severe clotting, and renal failure.

10. Shah, *The Fever*.

11. Maude, Woodrow, and White, "Artemisinin Antimalarials"; McNeil, "For Intrigue, Malaria Drug Gets the Prize."

12. Yearsley, "Artemisinin."

13. Maude, Woodrow, and White, "Artemisinin Antimalarials;" Yearsley, "Artemisinin."

14. Shah, *The Fever*.

15. Maude, Woodrow, and White, "Artemisinin Antimalarials"; McNeil, "For Intrigue, Malaria Drug Gets the Prize"; Shah, *The Fever*.

16. Clarke, "Malaria Is Killing One African Child Every 30 Seconds."

17. McNeil, "For Intrigue, Malaria Drug Gets the Prize"; Shah, *The Fever*.

18. Chorev, *World Health Organization between North and South*; Baxerres and Cassier, *Understanding Drugs Markets*.

19. Baxerres and Cassier, *Understanding Drugs Markets*.

20. Baxerres and Cassier.

21. Liu et al., "Global Research on Artemisinin and Its Derivatives."

22. Shah, *The Fever*.

23. McNeil, "For Intrigue, Malaria Drug Gets the Prize"; Shah, *The Fever*.

24. World Health Organization, *World Malaria Report 2021*; Jana et al., "Antimalarial Patent Landscape."

25. Liu et al., "Global Research on Artemisinin and Its Derivatives"; Chorev, *Give and Take*.

26. Buckley and Baker, "IMF Policies and Health in Sub-Saharan Africa"; Chorev, *Give and Take;* McMichael and Weber, *Development and Social Change*.

27. Chorev, *Give and Take*.

28. World Health Organization, *World Malaria Report 2019*.

29. Jorgensen, "Uganda."

30. Kajula et al., "Political Analysis of Rapid Change in Uganda's Health Financing Policy"; Sejjaaka, "Political and Economic History of Uganda."

31. Chorev, *World Health Organization between North and South*; Sejjaaka, "Political and Economic History of Uganda."

32. Okuonzi, "Dying for Economic Growth?"; Turshen, *Privatizing Health Services in Africa*.

33. Kajula et al., "Political Analysis of Rapid Change in Uganda's Health Financing Policy;" Kwesiga, Zikusooka, and Ataguba, "Assessing Catastrophic and Impoverishing Effects of Health Care Payments in Uganda."

34. Chorev, *Give and Take*; McKenzie and Samba, "Role of Mathematical Modeling in Evidence-Based Malaria Control."

35. Kajula et al., "Political Analysis of Rapid Change in Uganda's Health Financing Policy"; Chorev, *Give and Take*; Navarro, "Neoliberalism and Its Consequences."

36. Kajula et al., "Political Analysis of Rapid Change in Uganda's Health Financing Policy"; World Bank, "Current Health Expenditure per Capita (Current US$)—Uganda."

37. Buckley and Baker, "IMF Policies and Health in Sub-Saharan Africa"; Chandler et al., "Introducing Malaria Rapid Diagnostic Tests at Registered Drug Shops in Uganda."

38. Buckley and Baker, "IMF Policies and Health in Sub-Saharan Africa."

39. Chandler et al., "Introducing Malaria Rapid Diagnostic Tests at Registered Drug Shops in Uganda"; Okuonzi, "Dying for Economic Growth?"

40. Kajula et al., "Political Analysis of Rapid Change in Uganda's Health Financing Policy."

41. Chandler et al., "Introducing Malaria Rapid Diagnostic Tests at Registered Drug Shops in Uganda"; Espino and Manderson, "Treatment Seeking for Malaria in Morong, Bataan, the Philippines"; Mbonye, Magnussen, and Bygbjerg, "Intermittent Preventive Treatment of Malaria in Pregnancy;" Fritze et al., "Hope, Despair, and Transformation."

42. Chandler et al., "Introducing Malaria Rapid Diagnostic Tests at Registered Drug Shops in Uganda"; Espino and Manderson, "Treatment Seeking for Malaria in Morong, Bataan, the Philippines"; Mbonye, Magnussen, and Bygbjerg, "Intermittent Preventive Treatment of Malaria in Pregnancy"; Fritze et al., "Hope, Despair and Transformation."

43. Global Malaria Programme, "Artemisinin Resistance and Artemisinin-Based Combination Therapy Efficacy."

44. Yearsley, "Artemisinin."

45. Liu et al., "Global Research on Artemisinin and Its Derivatives."

46. Pawar, "It Took 35 Years to Get a Malaria Vaccine."

47. World Health Organization, *World Malaria Report 2021.*

48. Pawar, "It Took 35 Years to Get a Malaria Vaccine."

49. Pawar.

50. Jarry, "Malaria Vaccine's Success Story Hides Legitimate Concerns"; World Health Organization, *World Malaria Report 2021.*

51. World Health Organization, *World Malaria Report 2021.*

## CHAPTER 7. THE BANK OF THE PLANET

1. Krugman, "Rule by Rentiers."

2. Wolf, "Why Rigged Capitalism Is Damaging Liberal Democracy"; Wolf, "Wipe Out Rentiers with Cheap Money."

3. Robinson, *The Ministry for the Future.*

4. Christophers, "The Problem of Rent." Christophers's work adeptly critiques the use of the term *financialization* as encompassing an overly broad swath of financial and other social processes. It its place, he uses the terms *rent, rentier,* and *rentierism* to more specifically capture how profits from monopoly

ownership can be made from assets that are necessarily held by the financial sector. While we recognize and appreciate the analytical distinctions Christophers makes, we believe that term *finance* and its associated *-iers*, *-isms*, and *-izations* are more recognizable and intuitively understood by the everyday reader. We thus use the terms *rent*, *rentiers*, and *rentierism* interchangeably with *finance*, *financiers*, and *financialization*.

5. Shiller, *Finance and the Good Society.*

6. Hossein, *Politicized Microfinance*; Hockett, "Finance without Financiers."

7. Block, "Financial Democratization and the Transition to Socialism"; McCarthy, "Politics of Democratizing Finance."

8. Hockett, "Finance without Financiers."

9. Polanyi, *Great Transformation.*

10. Zhang, Hu, and Ji, "Financial Markets under the Global Pandemic of COVID-19"; Marin and Corsetti, "The Dollar and International Capital Flows in the COVID-19 Crisis."

11. Hale et al., "A Global Panel Database of Pandemic Policies"; Jasanoff et al., "Comparative Covid Response."

12. Block, "Financial Democratization and the Transition to Socialism."

13. For discussions of the weakening of WTO, see Hopewell, *Clash of Powers*; Hopewell, *Breaking the WTO.*

14. Prasad, *Gaining Currency*; Fiallo Flor, "Lula revive el fracasado proyecto chavista de moneda única en la region"; Khan, Sikarwar, and Bhat, "BRICS."

15. Jenkins, *Bonds of Inequality*; Hudson, *Bankers and Empire*; Hildyard, *Licensed Larceny.*

16. Baines and Hager, "From Passive Owners to Planet Savers?"; Fancy, "Secret Diary of a 'Sustainable Investor'—Part 1;" Goldstein, *Planetary Improvement.*

17. Gallagher and Myers, "China–Latin America Finance Database"; Parks, Malik, and Wooley, "Is Beijing a Predatory Lender?"

18. Federal Housing Finance Agency, "Conforming Loan Limit Values."

19. Robbins, *Lawn People.*

20. Gallagher and Myers, "China-Latin America Finance Database."

21. Oliveira, "New Actors in China-Brazil Financial Geography."

22. World Bank, "WBG Finances—Country Details—Brazil"; Gallagher and Myers, "China–Latin America Finance Database."

23. For a discussion of the changing roles of Chinese finance in Brazil, see Oliveira, "New Actors in China-Brazil Financial Geography."

24. Parks, Malik, and Wooley, "Is Beijing a Predatory Lender?"

25. Edwards, "Triumph of Biotechnology and Private Capital."

26. Schulthess et al., "Relative Contributions of NIH and Private Sector Funding"; Sampat and Lichtenberg, "What Are the Respective Roles of the Public and Private Sectors in Pharmaceutical Innovation?"

27. Dolgin, "Tangled History of mRNA Vaccines."

28. Keestra, "Structural Violence and the Biomedical Innovation System"; Nelsen, "Role of University Technology Transfer Operations in Assuring Access to Medicines and Vaccines"; Bajaj, Maki, and Stanford, "Vaccine Apartheid."

29. Litewka and Heitman, "Latin American Healthcare Systems in Times of Pandemic"; Ridde, Queuille, and Ndour, "Nine Misconceptions about Free Healthcare in Sub-Saharan Africa."

30. Lane and Kim, "Government Health Expenditure and Public Health Outcomes"; Montagu et al., "Private versus Public Strategies for Health Service Provision"; McGough et al., "How Does Health Spending in the U.S. Compare to Other Countries?"

31. Lam, "What Motivates Academic Scientists to Engage in Research Commercialization."

32. Gold et al., "Are Patents Impeding Medical Care and Innovation?"; Boldrin and Levine, "The Case against Patents."

33. Robinson, *Ministry for the Future.*

34. Von Hayek, *Denationalisation of Money*; Lenin, "Imperialism and the Split in Socialism"; Keynes, *General Theory of Employment, Interest, and Money,* 334.

35. Christophers, "The Problem of Rent."

36. Robinson, *Ministry for the Future.*

37. Proudhon, *Solution of the Social Problem.*

# Bibliography

Aalbers, Manuel B. "Financial Geography III: The Financialization of the City." *Progress in Human Geography* 44, no. 3 (2020): 595–607. https://doi.org/10.1177/0309132519853922.

Abreu, Filipe Vieira Santos de, Ieda Pereira Ribeiro, Anielly Ferreira-de-Brito, Alexandre Araujo Cunha dos Santos, Rafaella Moraes de Miranda, Iule de Souza Bonelly, Maycon Sebastião Alberto Santos Neves, et al. "*Haemagogus leucocelaenus* and *Haemagogus janthinomys* Are the Primary Vectors in the Major Yellow Fever Outbreak in Brazil, 2016–2018." *Emerging Microbes and Infections* 8, no. 1 (2019): 218–31. https://doi.org/10.1080/22221751.2019.1568180.

Adenle, Ademola A., Karin Wedig, and Hossein Azadi. "Sustainable Agriculture and Food Security in Africa: The Role of Innovative Technologies and International Organizations." *Technology in Society* 58 (2019): 101143. https://doi.org/10.1016/j.techsoc.2019.05.007.

Admati, Anat. "In Banking, It's All Other People's Money." *Washington Post*, April 22, 2016. https://www.washingtonpost.com/news/in-theory/wp/2016/04/22/weve-let-a-culture-of-deceit-and-fraud-ruin-our-economy/.

Agency for Healthcare Research and Quality. "Clinical Guidelines and Recommendations." 2021. https://www.ahrq.gov/prevention/guidelines/index.html.

Agency for Healthcare Research and Quality. "Evidence-Based Practice Centers." 2021. https://effectivehealthcare.ahrq.gov/about/epc.

Agency for Healthcare Research and Quality. "Medical Expenditure Panel Survey, Table II.B.2.b(1): Percent of private-sector enrollees that are enrolled in self-insured plans at establishments that offer health insurance by firm size and State: United States, 2020."." 2020. https://meps.ahrq.gov/data_stats /summ_tables/insr/state/series_2/2020/tiib2b1.pdf.

Agency for Healthcare Research and Quality. "Medical Expenditure Panel Survey Data Tools." Updated November 23, 2022. https://www.meps.ahrq .gov/data_stats/data_tools.jsp.

Ali, S. Harris, Creighton Connolly, and Roger Keil. *Pandemic Urbanism: Infectious Diseases on a Planet of Cities.* John Wiley & Sons, 2022.

Allan, Brian F., Felicia Keesing, and Richard S. Ostfeld. "Effect of Forest Fragmentation on Lyme Disease Risk." *Conservation Biology* 17, no. 1 (February 1, 2003): 267–72.

Alencar, Jeronimo, Cecilia Ferreira de Mello, Fernanda Morone, Hermano Gomes Albuquerque, Nicolau Maués Serra-Freire, Raquel M. Gleiser, Shayenne Olsson Freitas Silva, and Anthony Érico Guimarães. "Distribution of Haemagogus and Sabethes species in Relation to Forest Cover and Climatic Factors in the Chapada Dos Guimarães National Park, State of Mato Grosso, Brazil." *Journal of the American Mosquito Control Association* 34, no. 2 (2018): 85–92. https://doi.org/10.2987/18-6739.1.

Alphey, Luke, Mark Benedict, Romeo Bellini, Gary G. Clark, David A. Dame, Mike W. Service, and Stephen L. Dobson. "Sterile-Insect Methods for Control of Mosquito-Borne Diseases: An Analysis." *Vector-Borne and Zoonotic Diseases* 10, no. 3 (April 2010): 295–311.

Altizer, Sonia, Richard S. Ostfeld, Pieter T. J. Johnson, Susan Kutz, and C. Drew Harvell. "Climate Change and Infectious Diseases: From Evidence to a Predictive Framework." *Science* 341, no. 6145 (August 2, 2013): 514–19.

Alto, Barry W., L. Philip Lounibos, Stephen Higgs, and Steven A. Juliano. "Larval Competition Differentially Affects Arbovirus Infection in Aedes Mosquitoes." *Ecology* 86, no. 12 (December 2005): 3279–88.

Aminov, Rustam I. "A Brief History of the Antibiotic Era: Lessons Learned and Challenges for the Future." *Frontiers of Microbiology* 1, no. 134 (December 2010). https://doi.org/10.3389/fmicb.2010.00134.

Antoniou, Evangelia, Eirini Orovou, Angeliki Sarella, Maria Iliadou, Nikolaos Rigas, Ermioni Palaska, Georgios Iatrakis, and Maria Dagla. "Zika Virus and the Risk of Developing Microcephaly in Infants: A Systematic Review." *International Journal of Environmental Research and Public Health* 17, no. 11 (January 2020): 3806.

Antony, Hiasindh Ashmi, and Subhash Chandra Parija. "Antimalarial Drug Resistance: An Overview." *Tropical Parasitology* 6, no. 1 (2016): 30–41.

Araújo, Helena R. C., Danilo O. Carvalho, Rafaella S. Ioshino, André L. Costa-da-Silva, and Margareth L. Capurro. "*Aedes aegypti* Control Strategies in Brazil: Incorporation of New Technologies to Overcome the Persistence of Dengue Epidemics." *Insects* 6, no. 2 (June 11, 2015): 576–94.

Aronowitz, Robert A. "Lyme Disease: The Social Construction of a New Disease and Its Social Consequences." *Milbank Quarterly* 69, no. 1 (1991): 79–112.

Arrighi, Giovanni. *The Long Twentieth Century: Money, Power, and the Origins of Our Times.* 2nd ed. Verso Books, 2010.

Asidi, Alex, Raphael N'Guessan, Martin Akogbeto, Chris Curtis, and Mark Rowland. "Loss of Household Protection from Use of Insecticide-Treated Nets Against Pyrethroid-Resistant Mosquitoes, Benin." *Emerging Infectious Diseases* 18, no. 7 (July 2012): 1101–6.

Asiimwe, Godfrey B. "From Monopoly Marketing to Coffee Magendo: Responses to Policy Recklessness and Extraction in Uganda, 1971–79." *Journal of Eastern African Studies* 7, no. 1 (February 1, 2013): 104–24.

Baber, Walter F. "Privatizing Public Management: The Grace Commission and Its Critics." *Proceedings of the Academy of Political Science* 36, no. 3 (1987): 153–63.

Baer, Werner. *The Brazilian Economy: Growth and Development.* Lynne Rienner, 2008.

Baer, Werner, and Paul Beckerman. "The Trouble with Index-Linking: Reflections on the Recent Brazilian Experience." *World Development* 8, no. 9 (September 1, 1980): 677–703.

Baer, Werner, and Annibal V. Villela. "The Changing Nature of Development Banking in Brazil." *Journal of Interamerican Studies and World Affairs* 22, no. 4 (1980): 423–40.

Baffes, John. *Restructuring Uganda's Coffee Industry: Why Going Back to the Basics Matters.* Policy Research Working Papers. World Bank, 2006.

Baffes, John. "The Cotton Sector of Uganda: Africa Region Working Paper Series No. 123." World Bank, March 2009.

Baines, Joseph, and Sandy Brian Hager. "From Passive Owners to Planet Savers? Asset Managers, Carbon Majors, and the Limits of Sustainable Finance." *Competition and Change* 27, no. 3–4 (2023): 449–71.

Bajaj, Simar Singh, Lwando Maki, and Fatima Cody Stanford. "Vaccine Apartheid: Global Cooperation and Equity." *The Lancet* 399, no. 10334 (April 16, 2022): 1452–53. https://doi.org/10.1016/ S0140-6736(22)00328-2.

Barbour, Alan. *Lyme Disease: Why It's Spreading, How It Makes You Sick, and What to Do About It.* Johns Hopkins University Press, 2015.

Barcia, Manuel. *The Yellow Demon of Fever.* Yale University Press, 2020.

Barredo, Elina, and Matthew DeGennaro. "Not Just from Blood: Mosquito Nutrient Acquisition from Nectar Sources." *Trends in Parasitology* 36, no. 5 (May 1, 2020): 473–84. https://doi.org/10.1016/j.pt.2020.02.003.

Barros, Fábio S. M., and Nildimar A. Honório. "Deforestation and Malaria on the Amazon Frontier: Larval Clustering of *Anopheles darlingi* (Diptera: Culicidae) Determines Focal Distribution of Malaria." *American Journal of Tropical Medicine and Hygiene* 93, no. 5 (November 4, 2015): 939–53. https://doi.org/10.4269/ajtmh.15-0042.

Baxerres, Carine, and Maurice Cassier, eds. *Understanding Drugs Markets: An Analysis of Medicines, Regulations, and Pharmaceutical Systems in the Global South.* Routledge Studies in the Sociology of Health and Illness. Routledge, 2022.

Beaubien, Jason. "This Mosquito Likes Us Too Much for Our Own Good." *Goats and Soda: Stories in Life Changing World*, February 10, 2016. NPR. https://www.npr.org/sections/goatsandsoda/2016/02/10/466268138/this-mosquito-likes-us-too-much-for-our-own-good.

Benchimol, Jaime. "Yellow Fever Vaccine in Brazil: Fighting a Tropical Scourge, Modernising the Nation." In *The Politics of Vaccination*, edited by Christine Holmberg, Stuart Blume, and Paul Greenough. Manchester University Press, 2017. https://doi.org/10.7228/manchester/9781526110886.003.0008.

Benton, Mary Lauren, Abin Abraham, Abigail L. LaBella, Patrick Abbot, Antonis Rokas, and John A. Capra. "The Influence of Evolutionary History on Human Health and Disease." *Nature Reviews Genetics* 22, no. 5 (May 2021): 269–83. https://doi.org/10.1038/s41576-020-00305-9.

Berger, B. W., R. C. Johnson, C. Kodner, and L. Coleman. "Cultivation of *Borrelia burgdorferi* from Erythema Migrans Lesions and Perilesional Skin." *Journal of Clinical Microbiology* 30, no. 2 (February 1992): 359–61. https://doi.org/10.1128/jcm.30.2.359-361.1992.

Berger, Kathryn A., Howard S. Ginsberg, Katherine D. Dugas, Lutz H. Hamel, and Thomas N. Mather. "Adverse Moisture Events Predict Seasonal Abundance of Lyme Disease Vector Ticks (*Ixodes scapularis*)." *Parasites and Vectors* 7 (April 14, 2014): 181. https://doi.org/10.1186/1756-3305-7-181.

Bill & Melinda Gates Foundation. "About." Accessed September 15, 2023. https://www.gatesfoundation.org/about.

Bill & Melinda Gates Foundation. "About Grand Challenges." Accessed September 14, 2023. https://gcgh.grandchallenges.org/about.

Bill & Melinda Gates Foundation. "Bill and Melinda Gates Call for New Global Commitment to Chart a Course for Malaria Eradication." October 2007. https://www.gatesfoundation.org/ideas/media-center/press-releases/2007/10/chart-a-course-for-malaria-eradication.

Bill & Melinda Gates Foundation. "Foundation Fact Sheet." Accessed November 5, 2024. https://www.gatesfoundation.org/about/foundation-fact-sheet.

Bill & Melinda Gates Foundation. "Gates Foundation Commits $258.3 Million for Malaria Research and Development." October 2005. https://www

.gatesfoundation.org/ideas/media-center/press-releases/2005/10/gates
-foundation-commits-2583-million-for-malaria-research.

Bishopp, F. C., and Helen Louise Trembley. "Distribution and Hosts of Certain North American Ticks." *Journal of Parasitology* 31, no. 1 (1945): 1–54. https://doi.org/10.2307/3273061.

Block, Fred. "Financial Democratization and the Transition to Socialism." *Politics and Society* 47, no. 4 (December 1, 2019): 529–56. https://doi.org/10.1177/0032329219879274.

Block, Fred, and Robert Hockett. *Democratizing Finance: Restructuring Credit to Transform Society.* Verso Books, 2022.

Boldrin, Michele, and David K. Levine. "The Case Against Patents." *Journal of Economic Perspectives* 27, no. 1 (February 2013): 3–22. https://doi.org/10.1257/jep.27.1.3.

Bolen, Rebecca, and J. Camille Hall. "Managed Care and Evidence-Based Practice: The Untold Story." *Journal of Social Work Education* 43 (September 1, 2007): 463–79. https://doi.org/10.5175/JSWE.2007.200600656.

Bosco Siqueira, João, Celina Maria Turchi Martelli, Giovanini Evelim Coelho, Ana Cristina da Rocha Simplício, and Douglas L. Hatch. "Dengue and Dengue Hemorrhagic Fever, Brazil, 1981–2002." *Emerging Infectious Diseases* 11, no. 1 (January 2005): 48–53. https://doi.org/10.3201/eid1101.031091.

Brady, Oliver J., Nick Golding, David M. Pigott, Moritz U. G. Kraemer, Jane P. Messina, Robert C. Reiner Jr., Thomas W. Scott, David L. Smith, Peter W. Gething, and Simon I. Hay. "Global Temperature Constraints on *Aedes aegypti* and *Ae. albopictus* Persistence and Competence for Dengue Virus Transmission." *Parasites and Vectors* 7, no. 1 (July 22, 2014): 338. https://doi.org/10.1186/1756-3305-7-338.

Braga, Ima Aparecida, and Denise Valle. "*Aedes aegypti*: Histórico do Controle no Brasil." *Epidemiologia e Serviços de Saúde* 16, no. 2 (June 2007): 113–18. https://doi.org/10.5123/S1679-49742007000200006.

Brathwaite Dick, Olivia, José L. San Martín, Romeo H. Montoya, Jorge del Diego, Betzana Zambrano, and Gustavo H. Dayan. "The History of Dengue Outbreaks in the Americas." *American Journal of Tropical Medicine and Hygiene* 87, no. 4 (2012): 584–93. https://doi.org/10.4269/ajtmh.2012.11-0770.

Brook, R. H. "Practice Guidelines and Practicing Medicine: Are They Compatible?" *Journal of the American Medical Association* 262, no. 21 (December 1, 1989): 3027–30.

Brown, Julia E., Benjamin R. Evans, Wei Zheng, Vanessa Obas, Laura Barrera-Martinez, Andrea Egizi, Hongyu Zhao, Adalgisa Caccone, and Jeffrey R. Powell. "Human Impacts Have Shaped Historical and Recent Evolution in *Aedes aegypti*, the Dengue and Yellow Fever Mosquito." *Evolution:*

*International Journal of Organic Evolution* 68, no. 2 (February 2014): 514–25. https://doi.org/10.1111/evo.12281.

Brown, Phil, Steve Kroll-Smith, and Valerie Gunter. "Knowledge, Citizens, and Organizations." In *Illness and the Environment: A Reader in Contested Medicine*, edited by Steve Kroll-Smith, Phil Brown, and Valerie Gunter, 9–25. New York University Press, 2000.

Brown, Phil, Rachel Morello-Frosch, and Stephen Zavestoski. *Contested Illnesses: Citizens, Science, and Health Social Movements*. University of California Press, 2011. https://muse.jhu.edu/book/44539.

Bruegmann, Robert. *Sprawl: A Compact History*. University of Chicago Press, 2006.

Bryant, Juliet E., Edward C. Holmes, and Alan D. T. Barrett. "Out of Africa: A Molecular Perspective on the Introduction of Yellow Fever Virus into the Americas." *PLoS Pathogens* 3, no. 5 (May 18, 2007): e75. https://doi.org/10.1371/journal.ppat.0030075.

Buckley, Ross P., and Jonathan Baker. "IMF Policies and Health in Sub-Saharan Africa." In *Global Health Governance: Crisis, Institutions, and Political Economy*, edited by Adrian Kay and Owain David Williams, 209–26. International Political Economy. London: Palgrave Macmillan UK, 2009. https://doi.org/10.1057/9780230249486_10.

Bunker, Stephen G. *Peasants Against the State: The Politics of Market Control in Bugisu, Uganda, 1900–1983*. University of Chicago Press, 1987.

Bunker, Stephen G. *Underdeveloping the Amazon: Extraction, Unequal Exchange, and the Failure of the Modern State*. University of Chicago Press, 1990.

*Business Week*. "A New Try for S-P." August 9, 1958.

*Business Week*. "Studebaker-Packard Bets Its All." September 6, 1958.

Calavita, Kitty, Henry N. Pontell, and Robert Tillman. *Big Money Crime: Fraud and Politics in the Savings and Loan Crisis*. University of California Press, 1999.

Carroll, Marjorie Smith, and Ross H. Arnett. "Private Health Insurance Plans in 1978 and 1979: A Review of Coverage, Enrollment, and Financial Experience." *Health Care Financing Review* 3, no. 1 (September 1981): 55–87.

Carvalho, Danilo O., Andrew R. McKemey, Luiza Garziera, Renaud Lacroix, Christl A. Donnelly, Luke Alphey, Aldo Malavasi, and Margareth L. Capurro. "Suppression of a Field Population of *Aedes aegypti* in Brazil by Sustained Release of Transgenic Male Mosquitoes." *PLoS Neglected Tropical Diseases* 9, no. 7 (July 2, 2015): e0003864. https://doi.org/10.1371/journal.pntd.0003864.

Castro-Santos, Luiz Antonio de. "Power, Ideology, and Public Health in Brazil, 1889–1930." PhD dissertation, Department of Sociology, Harvard University, 1987. https://www.proquest.com/docview/303559071?fromopenview=true&pq-origsite=gscholar.

Catarina Andrioli, Denise, Maria Assunta Busato, and Junir Antonio Lutinski. "Spatial and Temporal Distribution of Dengue in Brazil, 1990–2017." *PLoS ONE* 15, no. 2 (February 13, 2020): e0228346. https://doi.org/10.1371/journal.pone.0228346.

Cavanagh, Connor Joseph, and David Himmelfarb. "'Much in Blood and Money': Necropolitical Ecology on the Margins of the Uganda Protectorate." *Antipode* 47, no. 1 (2015): 55–73. https://doi.org/10.1111/anti.12093.

Centers for Disease Control and Prevention. "Dengue Clinical Case Management (DCCM)—Home," 2018. https://www.cdc.gov/dengue/training/cme/ccm/index.html.

Centers for Disease Control and Prevention. "Global Health—Newsroom—Yellow Fever." 2018. https://www.cdc.gov/globalhealth/newsroom/topics/yellowfever/index.html.

Centers for Disease Control and Prevention. "How Many People Get Lyme Disease?" January 13, 2021. https://www.cdc.gov/lyme/stats/humancases.html.

Centers for Disease Control and Prevention. "Lyme Disease Treatment." 2019. https://www.cdc.gov/lyme/treatment/index.html.

Centers for Disease Control and Prevention. "Preventing Tick Bites on People." September 30, 2019. https://www.cdc.gov/lyme/prev/on_people.html.

Centers for Disease Control and Prevention. "Preventing Ticks in the Yard." February 22, 2019. https://www.cdc.gov/lyme/prev/in_the_yard.html.

Centers for Disease Control and Prevention. "Signs and Symptoms of Lyme Disease." January 15, 2021. https://www.cdc.gov/lyme/signs_symptoms/index.html.

Chandler, Clare I. R., Rachel Hall-Clifford, Turinde Asaph, Magnussen Pascal, Siân Clarke, and Anthony K. Mbonye. "Introducing Malaria Rapid Diagnostic Tests at Registered Drug Shops in Uganda: Limitations of Diagnostic Testing in the Reality of Diagnosis." *Social Science and Medicine* 72, no. 6 (2011): 937–44. https://doi.org/10.1016/j.socscimed.2011.01.009.

Chaplin-Kramer, Rebecca, Richard P. Sharp, Lisa Mandle, Sarah Sim, Justin Johnson, Isabela Butnar, Llorenç Milà i Canals, et al. "Spatial Patterns of Agricultural Expansion Determine Impacts on Biodiversity and Carbon Storage." *Proceedings of the National Academy of Sciences* 112, no. 24 (2015): 7402–7. https://doi.org/10.1073/pnas.1406485112.

Charles, Sarah. "Consolidation in the Coffee Industry Is Only Just Getting Started." *Coffee Intelligence*, June 8, 2023. https://intelligence.coffee/2023/06/consolidation-in-the-coffee-industry/.

Chippaux, Jean-Philippe, and Alain Chippaux. "Yellow Fever in Africa and the Americas: A Historical and Epidemiological Perspective." *Journal of Venomous Animals and Toxins Including Tropical Diseases* 24 (2018): 20. https://doi.org/10.1186/s40409-018-0162-y.

Chorev, Nitsan. *Give and Take: Developmental Foreign Aid and the Pharmaceutical Industry in East Africa.* Princeton Studies in Global and Comparative Sociology. Princeton University Press, 2019.

Chorev, Nitsan. *The World Health Organization between North and South.* Cornell University Press, 2012.

Chouin-Carneiro, Thais, Mariana Rocha David, Fernanda de Bruycker Nogueira, Flavia Barreto dos Santos, and Ricardo Lourenço-de-Oliveira. "Zika Virus Transmission by Brazilian *Aedes aegypti* and *Aedes albopictus* Is Virus Dose and Temperature-Dependent." *PLoS Neglected Tropical Diseases* 14, no. 9 (September 8, 2020): e0008527. https://doi.org/10.1371/journal.pntd.0008527.

Christophers, Brett. "The Problem of Rent." *Critical Historical Studies* 6, no. 2 (September 1, 2019): 303–23. https://doi.org/10.1086/705396.

Clapson, Mark. "Suburban Paradox? Planners' Intentions and Residents' Preferences in Two New Towns of the 1960s: Reston, Virginia and Milton Keynes, England." *Planning Perspectives* 17, no. 2 (January 1, 2002): 145–62. https://doi.org/10.1080/02665430110111856.

Clark, Brett, and John Bellamy Foster. "Ecological Imperialism and the Global Metabolic Rift: Unequal Exchange and the Guano/Nitrates Trade." *International Journal of Comparative Sociology* 50, nos. 3–4 (June 1, 2009): 311–34. https://doi.org/10.1177/0020715209105144.

Clarke, Tom. "Malaria Is Killing One African Child Every 30 Seconds." *Nature,* April 25, 2003. https://doi.org/10.1038/news030421-12.

Conrad, Melissa D., and Philip J. Rosenthal. "Antimalarial Drug Resistance in Africa: The Calm before the Storm?" *Lancet Infectious Diseases* 19, no. 10 (October 1, 2019): e338–51. https://doi.org/10.1016/S1473-3099(19)30261-0.

Cooper, D. B. "Brazil's Long Fight Against Epidemic Disease, 1849–1917, with Special Emphasis on Yellow Fever." *Bulletin of the New York Academy of Medicine* 51, no. 5 (May 1975): 672–96.

Costa, Frederico, Ticiana Carvalho-Pereira, Mike Begon, Lee Riley, and James Childs. "Zoonotic and Vector-Borne Diseases in Urban Slums: Opportunities for Intervention." *Trends in Parasitology* 33, no. 9 (September 1, 2017): 660–62. https://doi.org/10.1016/j.pt.2017.05.010.

Credit Suisse Research Institute. "Global Wealth Report 2023." https://www.ubs.com/global/en/family-office-uhnw/reports/global-wealth-report-2023.html.

Cronon, William. *Changes in the Land: Indians, Colonists, and the Ecology of New England.* Macmillan, 1983.

Daniels, Tom. *When City and Country Collide: Managing Growth in the Metropolitan Fringe.* Island Press, 1999.

Daviron, Benoit, and Stefano Ponte. *The Coffee Paradox: Global Markets, Commodity Trade, and the Elusive Promise of Development.* Zed Books, 2005. https://doi.org/10.5040/9781350222984.

Davis, Gerald F., and Suntae Kim. "Financialization of the Economy." *Annual Review of Sociology* 41, no. 1 (2015): 203–21. https://doi.org/10.1146/annurev-soc-073014-112402.

Davis, Mike. *The Monster at Our Door: The Global Threat of Avian Flu.* The New Press, 2005.

Davison, Graeme. "The Suburban Idea and Its Enemies." *Journal of Urban History* 39, no. 5 (September 1, 2013): 829–47. https://doi.org/10.1177/0096144213479307.

De Haas, Michiel. "The Failure of Cotton Imperialism in Africa: Seasonal Constraints and Contrasting Outcomes in French West Africa and British Uganda." *Journal of Economic History* 81, no. 4 (2021): 1098–1136. https://doi.org/10.1017/S0022050721000462.

De Haas, Michiel. "Measuring Rural Welfare in Colonial Africa: Did Uganda's Smallholders Thrive?" *Economic History Review* 70, no. 2 (2017): 605–31. https://doi.org/10.1111/ehr.12377.

De Haas, Michiel, and Kostadis J. Papaioannou. "Resource Endowments and Agricultural Commercialization in Colonial Africa: Did Labour Seasonality and Food Security Drive Uganda's Cotton Revolution?" EHES Working Papers in Economic History 111. European Historical Economics Society, 2017. Econstor. https://www.econstor.eu/handle/10419/247042.

Dean, Warren. *With Broadax and Firebrand: The Destruction of the Brazilian Atlantic Forest.* University of California Press, 1997.

Denevan, William M. "The Pristine Myth: The Landscape of the Americas in 1492." *Annals of the Association of American Geographers* 82, no. 3 (1992): 369–85.

Denning, Steve. "Why Financialization Has Run Amok." *Forbes*, June 3, 2014. https://www.forbes.com/sites/stevedenning/2014/06/03/why-financialization-has-run-amok/.

Denny, Keith. "Evidence-Based Medicine and Medical Authority." *Journal of Medical Humanities* 20, no. 4 (1999): 247–63.

Dickson, James G. "Wildlife of Southern Forests Habitat and Management: Early History." In *Wildlife of Southern Forests: Habitat and Management,* compiled and edited by James G. Dickson, 20–30. Hancock House, 2003. https://www.fs.usda.gov/treesearch/pubs/6966.

Dolgin, Elie. "The Tangled History of mRNA Vaccines." *Nature* 597, no. 7876 (September 14, 2021): 318–24. https://doi.org/10.1038/d41586-021-02483-w.

Dorit, Robert L. "Zika Goes Viral." *American Scientist, Research Triangle Park* 104, no. 5 (October 2016): 274–77.

Duca, John V., and Mark Walker. "Why Has U.S. Stock Ownership Doubled Since the Early 1980s? Equity Participation over the Past Half Century." *Federal Reserve Bank of Dallas, Working Papers 2022*, no. 2222 (November 2022). https://doi.org/10.24149/wp2222.

Dumes, Abigail A. *Divided Bodies: Lyme Disease, Contested Illness, and Evidence-Based Medicine.* Critical Global Health: Evidence, Efficacy, Ethnography. Duke University Press, 2020.

Edwards, Chris. "The Triumph of Biotechnology and Private Capital." Cato Institute, September 24, 2021. https://www.cato.org/commentary/triumph-biotechnology-private-capital.

Eisen, Rebecca J., and Lars Eisen. "The Blacklegged Tick, *Ixodes scapularis*: An Increasing Public Health Concern." *Trends in Parasitology* 34, no. 4 (April 2018): 295–309. https://doi.org/10.1016/j.pt.2017.12.006.

Eisen, Rebecca J., Lars Eisen, and Charles B. Beard. "County-Scale Distribution of *Ixodes scapularis* and *Ixodes pacificus* (Acari: Ixodidae) in the Continental United States." *Journal of Medical Entomology; Lanham* 53, no. 2 (March 2016): 349–86. http://dx.doi.org.proxy.wm.edu/10.1093/jme/tjv237.

Elbaum-Garfinkle, Shana. "Close to Home: A History of Yale and Lyme Disease." *Yale Journal of Biology and Medicine* 84, no. 2 (June 2011): 103–8.

Elias, Paulo Eduardo M., and Amelia Cohn. "Health Reform in Brazil: Lessons to Consider." *American Journal of Public Health* 93, no. 1 (January 2003): 44–48.

Emba, Christine. "Opinion: Has Our Economy Become Too 'Financialized'?" *Washington Post*, April 18, 2016. https://www.washingtonpost.com/news/in-theory/wp/2016/04/18/has-our-economy-become-too-financialized/.

Epstein, Gerald A. *Financialization and the World Economy.* Edward Elgar, 2005.

Espino, F., and L. Manderson. "Treatment Seeking for Malaria in Morong, Bataan, the Philippines." *Social Science and Medicine* 50, no. 9 (2000): 1309–16. https://doi.org/10.1016/s0277-9536(99)00379-2.

Espinosa, Mariola. "The Question of Racial Immunity to Yellow Fever in History and Historiography." *Social Science History* 38, no. 3–4 (ed 2014): 437–53. https://doi.org/10.1017/ssh.2015.20.

Estoque, Ronald C., Rajarshi Dasgupta, Karina Winkler, Valerio Avitabile, Brian A. Johnson, Soe W. Myint, Yan Gao, Makoto Ooba, Yuji Murayama, and Rodel D. Lasco. "Spatiotemporal Pattern of Global Forest Change over the Past 60 Years and the Forest Transition Theory." *Environmental Research Letters* 17, no. 8 (August 2022): 084022. https://doi.org/10.1088/1748-9326/ac7df5.

Fairbairn, Madeleine. *Fields of Gold: Financing the Global Land Rush.* Cornell University Press, 2020.

Fairbairn, Madeleine. "'Like Gold with Yield': Evolving Intersections Between Farmland and Finance." In *New Directions in Agrarian Political Economy*, edited by Ryan Isakson, 137–56. Routledge, 2015.

Fairfax County GIS and Mapping. "Fairfax County Historical Imagery Viewer." Accessed August 24, 2018. https://www.fairfaxcounty.gov/maps/aerial-photography.

Falcão Sobral, Marcos Felipe, and Ana Iza Gomes da Penha Sobral. "Casos de dengue e coleta de lixo urbano: um estudo na Cidade do Recife, Brasil." *Ciência e Saúde Coletiva* 24 (2019): 1075–82. https://doi.org/10.1590/1413-81232018243.10702017.

Fancy, Tariq. "The Secret Diary of a 'Sustainable Investor'—Part 1." *Medium* (blog), August 2021. https://medium.com/@sosofancy/the-secret-diary-of-a-sustainable-investor-part-1-70b6987fa139.

Federal Housing Finance Agency. "Conforming Loan Limit Values." 2022. https://www.fhfa.gov/DataTools/Downloads/Pages/Conforming-Loan-Limit.aspx.

Federal Housing Finance Agency. "National Average Contract Mortgage Rate History." 2017. https://www.fhfa.gov/DataTools/Downloads/Pages/National-Average-Contract-Mortgage-Rate-History.aspx#2017.

Federal Reserve. "Mortgage Debt Outstanding." Board of Governors of the Federal Reserve System, March 2020. https://www.federalreserve.gov/data/mortoutstand/current.htm.

Feinstein, Alvan R, and Ralph I Horwitz. "Problems in the 'Evidence' of 'Evidence-Based Medicine.'" *American Journal of Medicine* 103, no. 6 (December 1, 1997): 529–35. https://doi.org/10.1016/S0002-9343(97)00244-1.

Feldman Marzochi, Keyla Belízia. "Dengue in Brazil—Situation, Transmission, and Control: A Proposal for Ecological Control." *Memórias do Instituto Oswaldo Cruz* 89 (June 1994): 235–45. https://doi.org/10.1590/S0074-02761994000200023.

Ferreira, Flavio Devienne. "Inside the Mosquito Factory That Could Stop Dengue and Zika." *MIT Technology Review*, February 17, 2016. https://www.technologyreview.com/2016/02/17/162230/inside-the-mosquito-factory-that-could-stop-dengue-and-zika/.

Ferring, David, and Heidi Hausermann. "The Political Ecology of Landscape Change, Malaria, and Cumulative Vulnerability in Central Ghana's Gold Mining Country." *Annals of the American Association of Geographers* 109, no. 4 (July 4, 2019): 1074–91. https://doi.org/10.1080/24694452.2018.1535885.

Fiallo Flor, Mamela. "Lula revive el fracasado proyecto chavista de moneda única en la región." *PanAm Post*, May 2, 2022. https://panampost.com/mamela-fiallo/2022/05/01/lula-moneda-unica/.

Field, Marilyn J., and Kathleen N. Lohr, eds. *Clinical Practice Guidelines: Directions for a New Program*. National Academies Press, 1990.

Figueiredo, Mario L. G. de, Almério de C. Gomes, Alberto A. Amarilla, André de S. Leandro, Agnaldo de S. Orrico, Renato F. de Araujo, Jesuína do S. M. Castro, Edison L. Durigon, Victor H. Aquino, and Luiz T. M. Figueiredo. "Mosquitoes Infected with Dengue Viruses in Brazil." *Virology Journal* 7 (July 12, 2010): 152. https://doi.org/10.1186/1743-422X-7-152.

Florida, Richard L. "The Political Economy of Financial Deregulation and the Reorganization of Housing Finance in the United States." *International Journal of Urban and Regional Research* 10, no. 2 (June 1, 1986): 207–31. https://doi.org/10.1111/j.1468-2427.1986.tb00012.x.

Food and Agriculture Organization of the United Nations. "Statistical Yearbook: World Food and Agriculture 2021." 2021. https://www.fao.org/3/cb4477en/online/cb4477en.html.

Franco, Odair. *História da Febre-Amarela no Brasil*. Rio de Janeiro: Ministério da Saúde Departamento Nacional de Endemias Rurais, 1969.

Frank, Zephyr Lake. "Elite Families and Oligarchic Politics on the Brazilian Frontier: Mato Grosso, 1889–1937." *Latin American Research Review* 36, no. 1 (2001): 49–74.

Frieden, Jeffry A. *Banking on the World: The Politics of American International Finance*. Routledge, 1987.

Frieden, Jeffry A. "The Brazilian Borrowing Experience: From Miracle to Debacle and Back." *Latin American Research Review* 22, no. 1 (1987): 95–131.

Friedman, Andrew. *Covert Capital: Landscapes of Denial and the Making of U.S. Empire in the Suburbs of Northern Virginia*. University of California Press, 2013.

Frierson, J. Gordon. "The Yellow Fever Vaccine: A History." *Yale Journal of Biology and Medicine* 83, no. 2 (June 2010): 77–85.

Fritze, Jessica G., Grant A. Blashki, Susie Burke, and John Wiseman. "Hope, Despair, and Transformation: Climate Change and the Promotion of Mental Health and Wellbeing." *International Journal of Mental Health Systems* 2, no. 1 (2008): 13. https://doi.org/10.1186/1752-4458-2-13.

Frizon Rizzotto, Maria Lucia, and Gastão Wagner de Sousa Campos. "The World Bank and the Brazilian National Health System in the Beginning of the 21st Century." *Saúde e Sociedade* 25, no. 2 (June 2016): 263–76. https://doi.org/10.1590/S0104-12902016150960.

Fuentes Sarmiento, Patricio Egon. "The Spatial Distribution of Industrial Production in Brazil, 1960–1970." PhD dissertation, Victoria University of Manchester, 1983. https://www.proquest.com/dissertations-theses/spatial-distribution-industrial-production-brazil/docview/2410444774/se-2?accountid=15053.

Gabel, Jon. "Marketwatch: Ten Ways HMOs Have Changed During the 1990s." *Health Affairs* 16, no. 3 (May 1997): 134–45. https://doi.org/10.1377/hlthaff.16.3.134.

Gallagher, Kevin P., and Margaret Myers. "China–Latin America Finance Database." *The Dialogue*, 2022. https://www.thedialogue.org/map_list/.

Gallup Inc. "U.S. Stock Ownership Highest Since 2008." Gallup.com, May 24, 2023. https://news.gallup.com/poll/506303/stock-ownership-highest-2008.aspx.

Gandy, Matthew. "The Zoonotic City: Urban Political Ecology and the Pandemic Imaginary." *International Journal of Urban and Regional Research* 46, no. 2 (2022): 202–19. https://doi.org/10.1111/1468-2427.13080.

Gandy, Matthew. "Zoonotic Urbanization: Multispecies Urbanism and the Rescaling of Urban Epidemiology." *Urban Studies* 60, no. 13 (2023): 2529–49.

Garmany, Jeff. "Situating Fortaleza: Urban Space and Uneven Development in Northeastern Brazil." *Cities* 28, no. 1 (February 1, 2011): 45–52. https://doi.org/10.1016/j.cities.2010.08.004.

Garziera, Luiza, Michelle Cristine Pedrosa, Fabrício Almeida de Souza, Maylen Gómez, Márcia Bento Moreira, Jair Fernandes Virginio, Margareth Lara Capurro, and Danilo Oliveira Carvalho. "Effect of Interruption of Over-Flooding Releases of Transgenic Mosquitoes over Wild Population of *Aedes aegypti*: Two Case Studies in Brazil." *Entomologia Experimentalis et Applicata* 164, no. 3 (2017): 327–39. https://doi.org/10.1111/eea.12618.

Geddes, Barbara. "Building 'State' Autonomy in Brazil, 1930–1964." *Comparative Politics* 22, no. 2 (1990): 217–35. https://doi.org/10.2307/422315.

Gibb, Rory, David W. Redding, Kai Qing Chin, Christl A. Donnelly, Tim M. Blackburn, Tim Newbold, and Kate E. Jones. "Zoonotic Host Diversity Increases in Human-Dominated Ecosystems." *Nature* 584, no. 7821 (August 2020): 398–402. https://doi.org/10.1038/s41586-020-2562-8.

GIS Mapping Center, Arlington County, VA. "Historical Aerial Photographs." Accessed November 5, 2024. https://www.arcgis.com/home/webmap/viewer.html?webmap=28ea281cba6a4a5a8050df04c7fbb478.

Glaser, Vicki. "Interview with Luke Alphey, Ph.D." *Vector Borne and Zoonotic Diseases* 9, no. 2 (April 2009): 221–26. https://doi.org/10.1089/vbz.2009.3087.

Gleim, Elizabeth R., L. Mike Conner, Roy D. Berghaus, Michael L. Levin, Galina E. Zemtsova, and Michael J. Yabsley. "The Phenology of Ticks and the Effects of Long-Term Prescribed Burning on Tick Population Dynamics in Southwestern Georgia and Northwestern Florida." *PLoS ONE* 9, no. 11 (November 6, 2014): e112174. https://doi.org/10.1371/journal.pone.0112174.

Global Malaria Programme. "Artemisinin Resistance and Artemisinin-Based Combination Therapy Efficacy." World Health Organization, 2018. https://apps.who.int/iris/bitstream/handle/10665/274362/WHO-CDS-GMP-2018.18-eng.pdf.

Glunt, Katey D., Maureen Coetzee, Silvie Huijben, A. Alphonsine Koffi, Penelope A. Lynch, Raphael N'Guessan, Welbeck A. Oumbouke, Eleanore D. Sternberg, and Matthew B. Thomas. "Empirical and Theoretical Investigation into the Potential Impacts of Insecticide Resistance on the Effectiveness of Insecticide-Treated Bed Nets." *Evolutionary Applications* 11, no. 4 (2018): 431–41. https://doi.org/10.1111/eva.12574.

Goindin, Daniella, Christelle Delannay, Cédric Ramdini, Joël Gustave, and Florence Fouque. "Parity and Longevity of *Aedes aegypti* According to Temperatures in Controlled Conditions and Consequences on Dengue Transmission Risks." *PLoS ONE* 10 (August 10, 2015): e0135489. https://doi.org/10.1371/journal.pone.0135489.

Gold, E. Richard, Warren Kaplan, James Orbinski, Sarah Harland-Logan, and Sevil N-Marandi. "Are Patents Impeding Medical Care and Innovation?" *PLoS Medicine* 7, no. 1 (January 5, 2010): e1000208. https://doi.org/10.1371/journal.pmed.1000208.

Goldstein, Jesse. *Planetary Improvement: Cleantech Entrepreneurship and the Contradictions of Green Capitalism.* MIT Press, 2018.

Gotham, Kevin Fox. "The Secondary Circuit of Capital Reconsidered: Globalization and the U.S. Real Estate Sector." *American Journal of Sociology* 112, no. 1 (July 2006): 231–75. https://doi.org/10.1086/502695.

Gottschalk, Marie. *The Shadow Welfare State: Labor, Business, and the Politics of Health Care in the United States.* Cornell University Press, 2000.

Gottwalt, Allison. "Impacts of Deforestation on Vector-Borne Disease Incidence." *Columbia University Journal of Global Health* 3, no. 2 (November 3, 2013): 16–19. https://doi.org/10.7916/thejgh.v3i2.4864.

Grann, David. "Stalking Dr. Steere." *New York Times Magazine,* June 17, 2001. https://www.nytimes.com/2001/06/17/magazine/stalking-dr-steere.html.

Gray, Gary G. *Wildlife and People: The Human Dimensions of Wildlife Ecology.* University of Illinois Press, 1995.

Greeley, W. B. "The Relation of Geography to Timber Supply." *Economic Geography* 1, no. 1 (1925): 1–14. https://doi.org/10.2307/140095.

Guerra Nunes, Priscila Conrado, Regina Paiva Daumas, Juan Camilo Sánchez-Arcila, Rita Maria Ribeiro Nogueira, Marco Aurélio Pereira Horta, and Flávia Barreto dos Santos. "30 Years of Fatal Dengue Cases in Brazil: A Review." *BMC Public Health* 19, no. 1 (March 21, 2019): 329. https://doi.org/10.1186/s12889-019-6641-4.

Gwee, Sylvia Xiao Wei, Ashley L. St. John, Gregory C. Gray, and Junxiong Pang. "Animals as Potential Reservoirs for Dengue Transmission: A Systematic Review." *One Health* 12 (January 20, 2021): 100216. https://doi.org/10.1016/j.onehlt.2021.100216.

Gwyer, G. D. "Three International Commodity Agreements: The Experience of East Africa." *Economic Development and Cultural Change* 21, no. 3 (April 1973): 465–76. https://doi.org/10.1086/450647.

Hacker, Jacob S. "The Rise of Reform." Chapter 1 in *The Road to Nowhere: The Genesis of President Clinton's Plan for Health Security,* 10–41. Princeton University Press, 1997. https://doi.org/10.2307/j.ctvl73f17d.5.

Hale, Thomas, Noam Angrist, Rafael Goldszmidt, Beatriz Kira, Anna Petherick, Toby Phillips, Samuel Webster, et al. "A Global Panel Database of Pandemic Policies (Oxford COVID-19 Government Response Tracker)." *Nature Human Behaviour* 5, no. 4 (April 2021): 529–38. https://doi.org/10 .1038/s41562-021-01079-8.

Hall, Sarah. "Geographies of Money and Finance II: Financialization and Financial Subjects." *Progress in Human Geography* 36, no. 3 (June 1, 2012): 403–11. https://doi.org/10.1177/0309132511403889.

Han, Barbara A., Andrew M. Kramer, and John M. Drake. "Global Patterns of Zoonotic Disease in Mammals." *Trends in Parasitology* 32, no. 7 (July 1, 2016): 565–77. https://doi.org/10.1016/j.pt.2016.04.007.

Harris, Angela F., Andrew R. McKemey, Derric Nimmo, Zoe Curtis, Isaac Black, Siân A. Morgan, Marco Neira Oviedo, et al. "Successful Suppression of a Field Mosquito Population by Sustained Release of Engineered Male Mosquitoes." *Nature Biotechnology* 30, no. 9 (September 2012): 828–30. https://doi.org/10.1038/nbt.2350.

Harris, Angela F., Derric Nimmo, Andrew R. McKemey, Nick Kelly, Sarah Scaife, Christl A. Donnelly, Camilla Beech, William D. Petrie, and Luke Alphey. "Field Performance of Engineered Male Mosquitoes." *Nature Biotechnology* 29, no. 11 (November 2011): 1034–37. https://doi.org/10 .1038/nbt.2019.

Hayek, Friedrich A. von. *Denationalisation of Money: The Argument Refined; an Analysis of the Theory and Practice of Concurrent Currencies.* 3rd ed. Hobart Paper Special 70. Institute of Economic Affairs, 1990.

Heimberger, Tracey, Suzanne Jenkins, Harold Russell, and Richard Duma. "Epidemiology of Lyme Disease in Virginia." *American Journal of the Medical Sciences* 300, no. 5 (November 1, 1990): 283–87. https://doi.org /10.1097/00000441-199011000-00002.

Heimlich, Ralph E., and William D. Anderson. "Development at the Urban Fringe and Beyond: Impacts on Agriculture and Rural Land." Agricultural Economic Report No. 803. Economic Research Service, U.S. Department of Agriculture, 2001.

Helfand, Steven M. "The Political Economy of Agricultural Policy in Brazil: Decision Making and Influence from 1964 to 1992." *Latin American Research Review* 34, no. 2 (1999): 3–41.

Helfand, Steven M., and Gervásio Castro de Rezende. "The Impact of Sector-Specific and Economy-Wide Policy Reforms on the Agricultural Sector in Brazil, 1980–98." *Contemporary Economic Policy* 22, no. 2 (2004): 194–212.

Helleiner, Eric. *States and the Reemergence of Global Finance: From Bretton Woods to the 1990s*. Cornell University Press, 1996.

Helmuth, Laura. "Romney Targets Lyme Disease Conspiracy Theorists." *Slate*, September 29, 2012. https://slate.com/technology/2012/09/chronic-lyme-disease-delusion-romney-campaign-pushes-medical-nonsense.html.

Hemingway, Janet, Linda Field, and John Vontas. "An Overview of Insecticide Resistance." *Science* 298, no. 5591 (October 4, 2002): 96–97. https://doi.org/10.1126/science.1078052.

Henig, Jeffrey R. "Privatization in the United States: Theory and Practice." *Political Science Quarterly* 104, no. 4 (1989): 649–70. https://doi.org/10.2307/2151103.

Hildyard, Nicholas. *Licensed Larceny: Infrastructure, Financial Extraction, and the Global South*. Manchester University Press, 2016.

Hochman, Gilberto. "From Autonomy to Partial Alignment: National Malaria Programs in the Time of Global Eradication, Brazil, 1941–1961." *Canadian Bulletin of Medical History / Bulletin Canadien D'histoire De La Medecine* 25, no. 1 (2008): 161–92. https://doi.org/10.3138/cbmh.25.1.161.

Hochman, Gilberto. *The Sanitation of Brazil: Nation, State, and Public Health, 1889–1930*. University of Illinois Press, 2016.

Hockett, Robert C. "Finance without Financiers." *Politics and Society* 47, no. 4 (December 1, 2019): 491–527. https://doi.org/10.1177/0032329219882190.

Hoffmann, Susan. *Politics and Banking: Ideas, Public Policy, and the Creation of Financial Institutions*. Johns Hopkins University Press, 2001.

Hogan Lovells International, LLP. "Agreement for the Acquisition of the Entire Issued and to Be Issued Share Capital of Oxitec Limited." Securities and Exchange Commission. August 7, 2015. https://www.sec.gov/Archives/edgar/data/1356090/000119312515287266/d95323dex21.htm.

Holmes, Edward C., and S. Susanna Twiddy. "The Origin, Emergence, and Evolutionary Genetics of Dengue Virus." *Infection, Genetics and Evolution* 3, no. 1 (May 1, 2003): 19–28. https://doi.org/10.1016/S1567-1348(03)00004-2.

Holmgren, Torgny, Louis Kasekende, Michael Antingi-Ego, and Daniel Ddamu-lira. "Aid and Reform in Uganda." In *Aid in Africa*, edited by Devarajan Shantayanan, David R. Dollar, and Torgny Holmgren. World Bank, 2001.

Hopewell, Kristen. *Breaking the WTO: How Emerging Powers Disrupted the Neoliberal Project*. Stanford University Press, 2016.

Hopewell, Kristen. *Clash of Powers: US-China Rivalry in Global Trade Governance*. Cambridge University Press, 2020.

Hossein, Caroline Shenaz. *Politicized Microfinance: Money, Power, and Violence in the Black Americas*. University of Toronto Press, 2016.

Hudson, Michael. "We Can't Save the Economy Unless We Fix Our Debt Addiction." *Washington Post*, April 20, 2016. https://www.washingtonpost .com/news/in-theory/wp/2016/04/20/we-cant-save-the-economy-unless-we-fix-our-debt-addiction/.

Hudson, Peter James. *Bankers and Empire: How Wall Street Colonized the Caribbean*. University of Chicago Press, 2018.

Ilic, Irena, and Milena Ilic. "Global Patterns of Trends in Cholera Mortality." *Tropical Medicine and Infectious Disease* 8, no. 3 (March 13, 2023): 169. https://doi.org/10.3390/tropicalmed8030169.

Impieri Souza, Ariani, Marília Teixeira de Siqueira, Ana Laura Carneiro Gomes Ferreira, Clarice Umbelino de Freitas, Anselmo César Vasconcelos Bezerra, Adeylson Guimarães Ribeiro, and Adelaide Cássia Nardocci. "Geography of Microcephaly in the Zika Era: A Study of Newborn Distribution and Socio-environmental Indicators in Recife, Brazil, 2015–2016." *Public Health Reports* 133, no. 4 (July 1, 2018): 461–71. https://doi.org/10.1177 /0033354918777256.

Instituto Brasileiro de Geografia e Estatística. "Table 1288: Population in Demographic Censuses by Household Situation." Accessed August 9, 2021. https://sidra.ibge.gov.br/tabela/1288.

Instituto Brasileiro de Geografia e Estatística. "Censo Agropecuário." 2018. https://sidra.ibge.gov.br/pesquisa/censo-agropecuario/censo-agropecuario-2017.

Instituto Brasileiro de Geografia e Estatística. "Censo Demográfico," 2018. https://sidra.ibge.gov.br/pesquisa/censo-demografico/series-temporais /series-temporais/.

Instituto Nacional de Pesquisas Espaciais (INPE) Observação da Terra. "Monitoramento do Desmatamento da Floresta Amazônica Brasileira por Satélite." Accessed November 5, 2024. http://www.obt.inpe.br/OBT /assuntos/programas/amazonia/prodes.

Intercontinental Exchange. "Historical Monthly Volumes—Futures." ICE Report Center, 2024. https://www.ice.com/report/8.

International Monetary Fund, Fiscal Affairs Department. "2023 Global Debt Monitor." September 2023. www.imf.org.

International Trade Center. "Trade Map—List of Exporters for the Selected Product (Coffee, Whether or Not Roasted or Decaffeinated; Coffee Husks and Skins; Coffee Substitutes . . .)." ITC: Trade Map. Accessed March 17, 2024. https://www.trademap.org.

Intrexon. "Intrexon to Achieve $175M Cash Goal, Appoints Helen Sabzevari, PhD, as New President and CEO and Will Change Name to Precigen to Reflect Healthcare Focus." January 2, 2019. https://investors.precigen.com /news-releases/news-release-details/intrexon-achieve-175m-cash-goal-appoints-helen-sabzevari-phd-new/.

Ioannidis, John P. A. "Evidence-Based Medicine Has Been Hijacked: A Report to David Sackett." *Journal of Clinical Epidemiology* 73 (May 2016): 82–86. https://doi.org/10.1016/j.jclinepi.2016.02.012.

Ioris, Antonio Augusto Rossotto. *Agribusiness and the Neoliberal Food System in Brazil: Frontiers and Fissures of Agro-Neoliberalism*. Routledge, 2017.

Ioris, Rafael R. *Transforming Brazil: A History of National Development in the Postwar Era*. Routledge, 2014.

Isis Innovation [Oxford University Innovation]. "New Technique to Control Killer Disease Offered Multi Million Dollar Funding," June 29, 2005. http://www.isis-innovation.com/news/news/oxitec-jul05.html.

Isis Innovation [Oxford University Innovation]. "Starting a Spinout Company." 2016. https://innovation.ox.ac.uk/wp-content/uploads/2014/04/Spin-out-Researcher-Booklet-20160121.pdf.

Iyamulemye, Emmanuel. "The History of Coffee in Uganda." *African Fine Coffees Review Magazine*, September 2017.

Jackson, Kenneth T. *Crabgrass Frontier: The Suburbanization of the United States*. Oxford University Press, 1987.

Jana, Tarkanta, Siddhartha Dulakakhoria, Deepak Bindal, and Ankit Tripathi. "Antimalarial Patent Landscape: A Qualitative and Quantitative Analysis." *Current Science* (2012).

Jarry, Jonathan. "The Malaria Vaccine's Success Story Hides Legitimate Concerns." McGill Office for Science and Society, 2021. https://www.mcgill.ca/oss/article/health-and-nutrition/malaria-vaccines-success-story-hides-legitimate-concerns.

Jasanoff, Sheila, Stephen Hilgartner, J Benjamin Hurlbut, Onur Özgöde, and Margarita Rayzberg. "Comparative Covid Response: Crisis, Knowledge, Politics—Interim Report." John F. Kennedy School of Government, Harvard University, January 2021. https://compcore.cornell.edu/wp-content/uploads/2021/03/Comparative-Covid-Response_Crisis-Knowledge-Politics_Interim-Report.pdf.

Jecker, Nancy S., and Caesar A. Atuire. "What's Yours Is Ours: Waiving Intellectual Property Protections for COVID-19 Vaccines." *Journal of Medical Ethics* 47, no. 9 (September 2021): 595–98. https://doi.org/10.1136/medethics-2021-107555.

Jenkins, Destin. *The Bonds of Inequality: Debt and the Making of the American City*. University of Chicago Press, 2022.

Johns Hopkins Coronavirus Resource Center. "Mortality Analyses." Accessed September 21, 2023. https://coronavirus.jhu.edu/data/mortality.

Jones, Kate E., Nikkita G. Patel, Marc A. Levy, Adam Storeygard, Deborah Balk, John L. Gittleman, and Peter Daszak. "Global Trends in Emerging Infectious Diseases." *Nature* 451, no. 7181 (February 2008): 990–93. https://doi.org/10.1038/nature06536.

Jorgensen, Jan Jelmert. "Uganda: A Modern History." Edited by Will Kaberuka. *Journal of Modern African Studies* 20, no. 2 (1982): 345–47.

Jorgenson, Andrew K., Kelly Austin, and Christopher Dick. "Ecologically Unequal Exchange and the Resource Consumption / Environmental Degradation Paradox: A Panel Study of Less-Developed Countries, 1970–2000." *International Journal of Comparative Sociology* 50, no. 3–4 (June 1, 2009): 263–84. https://doi.org/10.1177/0020715209105142.

Kaberuka, Will. "Uganda's 1946 Colonial Development Plan: An Appraisal." *Transafrican Journal of History* 16 (1987): 185–205.

Kaiser Family Foundation. "Employer Health Benefits Survey—Section 5: Market Shares of Health Plans." *KFF* (blog), November 10, 2021. https://www.kff.org/report-section/ehbs-2021-section-5-market-shares-of-health-plans/.

Kajula, Peter Waalwo, Francis Kintu, John Barugahare, and Stella Neema. "Political Analysis of Rapid Change in Uganda's Health Financing Policy and Consequences on Service Delivery for Malaria Control." *International Journal of Health Planning and Management* 19, no. S1 (2004): S133–S153. https://doi.org/10.1002/hpm.772.

Kaup, Brent Z. "The Making of Lyme Disease: A Political Ecology of Ticks and Tick-Borne Illness in Virginia." *Environmental Sociology* 4, no. 3 (2018): 381–91. https://doi.org/10.1080/23251042.2018.1436892.

Kaup, Brent Z. "Pathogenic Metabolism: A Rift and the Zika Virus in Mato Grosso, Brazil." *Antipode: A Radical Journal of Geography* 53, no. 2 (March 2021): 567–86. https://doi.org/10.1111/anti.12694.

Kaup, Brent Z., Matthew Abel, and Amanda Sikirica. "Individualized Environments, Individual Cures: An Examination of Lyme Disease Activism in Virginia." *Environment and Planning E: Nature and Space* 4, no. 2 (June 1, 2021): 545–63. https://doi.org/10.1177/2514848620923593.

Keestra, Sarai. "Structural Violence and the Biomedical Innovation System: What Responsibility Do Universities Have in Ensuring Access to Health Technologies?" *BMJ Global Health* 6, no. 5 (May 2021): e004916. https://doi.org/10.1136/bmjgh-2020-004916.

Keil, Roger. "Extended Urbanization, 'Disjunct Fragments,' and Global Suburbanisms." *Environment and Planning D: Society and Space* 36, no. 3 (June 1, 2018): 494–511. https://doi.org/10.1177/0263775817749594.

Keil, Roger. *Suburban Planet: Making the World Urban from the Outside In.* John Wiley & Sons, 2017.

Keynes, John Maynard. *The General Theory of Employment, Interest, and Money.* 1936. Reprint, Springer, 2018.

Khan, Sameer Ahmad, Satyendra Singh Sikarwar, and Waseem Ahmad Bhat. "BRICS: An Alternative to the World Bank and IMF." *International Journal of Research and Analytical Reviews* 6, no. 1 (2019).

Kikuti, Mariana, Geraldo M. Cunha, Igor A. D. Paploski, Amelia M. Kasper, Monaise M. O. Silva, Aline S. Tavares, Jaqueline S. Cruz, et al. "Spatial Distribution of Dengue in a Brazilian Urban Slum Setting: Role of Socioeconomic Gradient in Disease Risk." *PLoS Neglected Tropical Diseases* 9, no. 7 (July 21, 2015). https://doi.org/10.1371/journal.pntd.0003937.

Kimball, Thomas, and Raymond Johnson. "The Richness of American Wildlife." In *Wildlife and America: Contributions to an Understanding of American Wildlife and Its Conservation*, 2–17. Council on Environmental Quality, 1978.

Kindhauser, Mary Kay, Tomas Allen, Veronika Frank, Ravi Shankar Santhana, and Christopher Dye. "Zika: The Origin and Spread of a Mosquito-Borne Virus." *Bulletin of the World Health Organization* 94 (February 9, 2016). https://doi.org/10.2471/BLT.16.171082.

King, Brian. *States of Disease: Political Environments and Human Health*. University of California Press, 2017.

Klassen, W., C. F. Curtis, and J. Hendrichs. "History of the Sterile Insect Technique." In *Sterile Insect Technique: Principles and Practice in Area-Wide Integrated Pest Management*, edited by V. A. Dyck, J. Hendrichs, and A. S. Robinson, 2nd Edition., 1–44. CRC Press, 2021.

Klink, Carlos Augusto, and Adriana G. Moreira. "Past and Current Human Occupation, and Land Use." In *The Cerrados of Brazil: Ecology and Natural History of a Neotropical Savanna*, edited by Paulo S. Oliveira and Robert J. Marquis, 69–88. Columbia University Press, 2002. https://doi.org/10.7312/oliv12042-004.

Klink, Carlos Augusto, Adriana G. Moreira, and Otto Thomas Solbrig. "Ecological Impact of Agricultural Development in the Brazilian Cerrados." In *The World's Savannas: Economic Driving Forces, Ecological Constraints, and Policy Options for Sustainable Land Use*, edited by M. D. Young and O. T. Solbrig, 259–82. Parthenon Publishing Group, 1993.

Koster, Martijn, and Monique Nuijten. "From Preamble to Post-project Frustrations: The Shaping of a Slum Upgrading Project in Recife, Brazil." *Antipode* 44, no. 1 (2012): 175–96. https://doi.org/10.1111/j.1467-8330.2011.00894.x.

Krippner, Greta R. *Capitalizing on Crisis*. Harvard University Press, 2011.

Krishna, Anirudh. *One Illness Away: Why People Become Poor and How They Escape Poverty*. Oxford University Press, 2011.

Krugman, Paul. "Rule by Rentiers." *New York Times*, June 10, 2011, sec. Opinion. https://www.nytimes.com/2011/06/10/opinion/10krugman.html.

Kugeler, Kiersten J., Grace M. Farley, John D. Forrester, and Paul S. Mead. "Geographic Distribution and Expansion of Human Lyme Disease, United States." *Emerging Infectious Diseases* 21, no. 8 (August 2015): 1455–57. https://doi.org/10.3201/eid2108.141878.

Kulikoff, Allan. *Tobacco and Slaves: The Development of Southern Cultures in the Chesapeake, 1680–1800.* University of North Carolina Press, 1986.

Kwashirai, Vimbai C. "Environmental Change, Control, and Management in Africa." *Global Environment* 6, no. 12 (January 1, 2013): 166–96. https://doi.org/10.3197/ge.2013.061208.

Kwesiga, Brendan, Charlotte M. Zikusooka, and John E. Ataguba. "Assessing Catastrophic and Impoverishing Effects of Health Care Payments in Uganda." *BMC Health Services Research* 15 (2015): 30. https://doi.org/10.1186/s12913-015-0682-x.

Lam, Alice. "What Motivates Academic Scientists to Engage in Research Commercialization: 'Gold,' 'Ribbon,' or 'Puzzle'?" *Research Policy* 40, no. 10 (December 1, 2011): 1354–68. https://doi.org/10.1016/j.respol.2011.09.002.

Laman, Mineke, Beatrice Khamati, and Patrick Milimo. "Mount Elgon Integrated Conservation and Development Project (MEICDP): External Evaluation Final Report." March 31, 2001. IUCN, posted May 2022. https://www.iucn.org/sites/default/files/2022-05/mt_elgon_eval_final.pdf.

Lane, Shannon, and T. K. Kim. "Government Health Expenditure and Public Health Outcomes." *American International Journal of Contemporary Research* 3 (January 1, 2013): 1–13.

Langley, Paul. *The Everyday Life of Global Finance: Saving and Borrowing in Anglo-America.* Oxford University Press, 2008.

Langley, Paul. "Securitising Suburbia: The Transformation of Anglo-American Mortgage Finance." *Competition and Change* 10, no. 3 (2006): 283-99. https://doi.org/10.1179/102452906X1143.

Lantos, Paul M., Jeffrey Rumbaugh, Linda K. Bockenstedt, Yngve T. Falck-Ytter, Maria E. Aguero-Rosenfeld, Paul G. Auwaerter, Kelly Baldwin, et al. "Clinical Practice Guidelines by the Infectious Diseases Society of America (IDSA), American Academy of Neurology (AAN), and American College of Rheumatology (ACR): 2020 Guidelines for the Prevention, Diagnosis, and Treatment of Lyme Disease." *Clinical Infectious Diseases* 72, no. 1 (January 1, 2021): e1–48. https://doi.org/10.1093/cid/ciaa1215.

Lapavitsas, Costas. "The Government Isn't to Blame for the Rise of Wall Street." *Washington Post,* April 19, 2016. https://www.washingtonpost.com/news/in-theory/wp/2016/04/19/the-government-isnt-to-blame-for-the-rise-of-wall-street/.

Laporta, Gabriel Zorello, Paulo Inácio Knegt Lopez de Prado, Roberto André Kraenkel, Renato Mendes Coutinho, and Maria Anice Mureb Sallum. "Biodiversity Can Help Prevent Malaria Outbreaks in Tropical Forests." *PLOS Neglected Tropical Diseases* 7, no. 3 (March 21, 2013): e2139. https://doi.org/10.1371/journal.pntd.0002139.

Lardner, James. "Are We Repeating History by Letting Our Financial Sector Get Too Big?" *Washington Post,* April 20, 2016. https://www.washingtonpost

.com/news/in-theory/wp/2016/04/20/are-we-repeating-history-by-letting
-our-financial-sector-grow-too-large/.

Lenin, V. I. "Imperialism and the Split in Socialism." *Sbornik Sotsial-Demokrata*, no. 2 (December 1916). Marxists Internet Archive. https://www.marxists.org/archive/lenin/works/1916/oct/x01.htm.

Leyshon, Andrew, and Nigel Thrift. "The Capitalization of Almost Everything: The Future of Finance and Capitalism." *Theory, Culture, and Society* 24, no. 7–8 (December 1, 2007): 97–115. https://doi.org/10.1177/0263276407084699.

Liang, Liucun, Jinyong Wang, Lucas Schorter, Thu Phong Nguyen Trong, Shari Fell, Sebastian Ulrich, and Reinhard K. Straubinger. "Rapid Clearance of *Borrelia burgdorferi* from the Blood Circulation." *Parasites and Vectors* 13, no. 1 (April 21, 2020): 191. https://doi.org/10.1186/s13071-020-04060-y.

Lima Amaral, Ernesto Friedrich de. "Brazil: Internal Migration." In *The Encyclopedia of Global Human Migration*, edited by Immanuel Ness. Blackwell, 2013. https://doi.org/10.1002/9781444351071.wbeghm075.

Lima, Andrew, Diane D. Lovin, Paul V. Hickner, and David W. Severson. "Evidence for an Overwintering Population of *Aedes aegypti* in Capitol Hill Neighborhood, Washington, DC." *American Journal of Tropical Medicine and Hygiene* 94, no. 1 (January 6, 2016): 231–35. https://doi.org/10.4269/ajtmh.15-0351.

Lima, Nísia Trindade. "Public Health and Social Ideas in Modern Brazil." *American Journal of Public Health* 97, no. 7 (July 2007): 1168–77. https://doi.org/10.2105/AJPH.2003.036020.

Litewka, Sergio G., and Elizabeth Heitman. "Latin American Healthcare Systems in Times of Pandemic." *Developing World Bioethics* 20, no. 2 (2020): 69–73. https://doi.org/10.1111/dewb.12262.

Liu, Kunmeng, Huali Zuo, Guoguo Li, Hua Yu, and Yuanjia Hu. "Global Research on Artemisinin and Its Derivatives: Perspectives from Patents." *Pharmacological Research* 159 (September 2020): 105048. https://doi.org/10.1016/j.phrs.2020.105048.

Lloyd-Sherlock, Peter. "The Recent Appearance of Favelas in São Paulo City: An Old Problem in a New Setting." *Bulletin of Latin American Research* 16, no. 3 (September 1, 1997): 289–305. https://doi.org/10.1016/S0261-3050(96)00030-7.

LoGiudice, Kathleen, Richard S. Ostfeld, Kenneth A. Schmidt, and Felicia Keesing. "The Ecology of Infectious Disease: Effects of Host Diversity and Community Composition on Lyme Disease Risk." *Proceedings of the National Academy of Sciences* 100, no. 2 (January 21, 2003): 567–71. https://doi.org/10.1073/pnas.0233733100.

Lowe, Rachel, Christovam Barcellos, Patrícia Brasil, Oswaldo G. Cruz, Nildimar Alves Honório, Hannah Kuper, and Marilia Sá Carvalho. "The Zika Virus Epidemic in Brazil: From Discovery to Future Implications." *Interna-*

*tional Journal of Environmental Research and Public Health* 15, no. 1 (January 2018): 96. https://doi.org/10.3390/ijerph15010096.

Löwy, Ilana. "Epidemiology, Immunology, and Yellow Fever: The Rockefeller Foundation in Brazil, 1923–1939." *Journal of the History of Biology* 30, no. 3 (1997): 397–417.

Löwy, Ilana. "Leaking Containers: Success and Failure in Controlling the Mosquito *Aedes aegypti* in Brazil." *American Journal of Public Health* 107 (2017): 517–24.

Lyme and Tick-Borne Diseases Research Center. "Diagnosis." Irving Medical Center, Columbia University. April 11, 2018. https://www.columbia-lyme .org/diagnosis.

Macinko, James, and Matthew J. Harris. "Brazil's Family Health Strategy: Delivering Community-Based Primary Care in a Universal Health System." *New England Journal of Medicine* 372, no. 23 (June 4, 2015): 2177–81. https://doi.org/10.1056/NEJMp1501140.

Macinko, James, and Maria Fernanda Lima-Costa. "Horizontal Equity in Health Care Utilization in Brazil, 1998–2008." *International Journal for Equity in Health* 11, no. 1 (June 21, 2012): 33. https://doi.org/10.1186 /1475-9276-11-33.

Mamdani, Mahmood. *Politics and Class Formation in Uganda*. Monthly Review Press, 1976.

Marin, Emile, and Giancarlo Corsetti. "The Dollar and International Capital Flows in the COVID-19 Crisis." *CEPR: VoxEU*, April 3, 2020. https://cepr .org/voxeu/columns/dollar-and-international-capital-flows-covid-19-crisis.

Marin, Jorge. "Empresa lança assinatura de 'mosquitos do bem' que combatem dengue." *Tecmundo,* November 15, 2021. https://www.tecmundo.com.br /ciencia/228689-empresa-lanca-assinatura-mosquitos-combatem-dengue .htm.

Marques, Marcia C. M., and Carlos E. V. Grelle, eds. *The Atlantic Forest: History, Biodiversity, Threats, and Opportunities of the Mega-Diverse Forest.* Springer, 2021.

Marques, Marcia C. M., Weverton Trindade, Amabily Bohn, and Carlos E. V. Grelle. "The Atlantic Forest: An Introduction to the Megadiverse Forest of South America." In *The Atlantic Forest: History, Biodiversity, Threats, and Opportunities of the Mega-diverse Forest,* edited by Marcia C. M. Marques and Carlos E. V. Grelle, 3–23. Cham, Switzerland: Springer International, 2021. https://doi.org/10.1007/978-3-030-55322-7_1.

Marshall, J. M. "The Cartagena Protocol in the Context of Recent Releases of Transgenic and Wolbachia-Infected Mosquitoes." *Asia-Pacific Journal of Molecular Biology and Biotechnology* 19 (January 1, 2013): 93–100.

Martins, Bruno, Eduardo Lundberg, and Tony Takeda. "Housing Finance in Brazil: Institutional Improvements and Recent Developments." SSRN

Scholarly Paper. Rochester, NY: Social Science Research Network, September 1, 2011. https://doi.org/10.2139/ssrn.1946143.

Martins, Helen. "Justiça autoriza empresa a comercializar *Aedes aegypti* modificado." *Agência Brasil,* March 23, 2018. https://agenciabrasil.ebc.com .br/pesquisa-e-inovacao/noticia/2018-03/justica-autoriza-empresa -comercializar-aedes-aegypti-modificado.

Marx, Karl. *The Eighteenth Brumaire of Louis Bonaparte.* 1852. Marxists Internet Archive. https://www.marxists.org/archive/marx/works/1852 /18th-brumaire/.

Masefield, G. B. "Agricultural Change in Uganda, 1945–60." *Journal of Modern African Studies* 1, no. 3 (1962): 416–17. https://doi.org/10.1017 /S0022278X00001944.

Masiga, Moses, and Alice Ruhweza. "Commodity Revenue Management: Coffee and Cotton in Uganda." IISD: International Institute for Sustainable Development, 2007. https://www.iisd.org/system/files/publications/trade_ price_case_coffee_cotton.pdf.

Mason, David L. *From Buildings and Loans to Bail-Outs: A History of the American Savings and Loan Industry, 1831–1995.* Cambridge University Press, 2004.

Maude, Richard J., Charles J. Woodrow, and Lisa J. White. "Artemisinin Antimalarials: Preserving the 'Magic Bullet.'" *Drug Development Research* 71, no. 1 (2010): 12–19. https://doi.org/10.1002/ddr.20344.

Mbonye, Anthony K., Pascal Magnussen, and I. B. Bygbjerg. "Intermittent Preventive Treatment of Malaria in Pregnancy: The Effect of New Delivery Approaches on Access and Compliance Rates in Uganda." *Tropical Medicine and International Health* 12, no. 4 (2007): 519–31. https://doi.org/10.1111 /j.1365-3156.2007.01819.x.

McBride, Carolyn S., Felix Baier, Aman B. Omondi, Sarabeth A. Spitzer, Joel Lutomiah, Rosemary Sang, Rickard Ignell, and Leslie B. Vosshall. "Evolution of Mosquito Preference for Humans Linked to an Odorant Receptor." *Nature* 515, no. 7526 (November 2014): 222–27. https://doi.org/10.1038 /nature13964.

McCabe, R. E., and R. T. McCabe. "Of Slings and Arrows: A Historical Perspective." In *White-Tailed Deer: Ecology and Management,* 19–72. Stackpole Books, 1984.

McCarthy, Michael A. "The Politics of Democratizing Finance: A Radical View." *Politics and Society* 47, no. 4 (December 1, 2019): 611–33. https://doi.org /10.1177/0032329219878990.

McDonald's. "100 Circle Farms." Accessed September 22, 2022. https://www .mcdonalds.com/us/en-us/about-our-food/meet-our-suppliers/100-circle -farms.html.

McGough, Matthew, Imani Telesford, Shameek Rakshit, Emma Wager, Krutika Amin, and Cynthia Cox. "How Does Health Spending in the U.S. Compare to Other Countries?" *Peterson-KFF Health System Tracker* (blog), February 9, 2023. https://www.healthsystemtracker.org/chart-collection/health-spending -u-s-compare-countries/.

McKenzie, F. Ellis, and Ebrahim M. Samba. "The Role of Mathematical Modeling in Evidence-Based Malaria Control." *American Journal of Tropical Medicine and Hygiene* 71, no. S2 (2004): S94–S96.

McMichael, Philip. *Development and Social Change: A Global Perspective.* 5th edition. Sage, 2012.

McMichael, Philip. "Incorporating Comparison within a World-Historical Perspective: An Alternative Comparative Method." *American Sociological Review* 55, no. 3 (1990): 385–97. https://doi.org/10.2307/2095763.

McMichael, Philip. "World-Systems Analysis, Globalization, and Incorporated Comparison." *Journal of World-Systems Research*, November 26, 2000, 668–89. https://doi.org/10.5195/jwsr.2000.192.

McMichael, Philip, and Heloise Weber. *Development and Social Change: A Global Perspective.* 7th edition. Sage, 2022.

McNeil, Donald G. "For Intrigue, Malaria Drug Gets the Prize." *New York Times,* January 16, 2012, sec. Health. https://www.nytimes.com/2012/01 /17/health/for-intrigue-malaria-drug-artemisinin-gets-the-prize.html.

McNeil, Donald G. "New Drug for Malaria Pits U.S. Against Africa." *New York Times,* May 28, 2002, sec. Health. https://www.nytimes.com/2002/05/28 /health/new-drug-for-malaria-pits-us-against-africa.html.

McNeill, John R. "Agriculture, Forests, and Ecological History: Brazil, 1500– 1984." *Environmental Review* 10, no. 2 (1986): 123–33. https://doi.org/10 .2307/3984562.

McNeill, John R. "Ecology, Epidemics, and Empires: Environmental Change and the Geopolitics of Tropical America, 1600–1825." *Environment and History* 5, no. 2 (1999): 175–84.

McNeill, John R. *Mosquito Empires: Ecology and War in the Greater Carib-bean, 1620–1914.* Cambridge University Press, 2010.

McNeill, William H. *Plagues and Peoples.* Anchor Books, 1976.

McPherson, Michelle, Almudena García-García, Francisco José Cuesta-Valero, Hugo Beltrami, Patti Hansen-Ketchum, Donna MacDougall, and Nicholas Ogden. "Expansion of the Lyme Disease Vector *Ixodes scapularis* in Canada Inferred from CMIP5 Climate Projections." *Environmental Health Perspectives* 125, no. 5 (May 31, 2017): 057008. https://doi.org/10.1289 /EHP57.

Mead, Paul, Jeannine Peterson, and Alison Hinckley. "Updated CDC Recom-mendation for Serologic Diagnosis of Lyme Disease." *MMWR: Morbidity*

*and Mortality Weekly Report* 68 (2019). https://doi.org/10.15585/mmwr
.mm6832a4.

Mechanic, David. "The Managed Care Backlash: Perceptions and Rhetoric in
Health Care Policy and the Potential for Health Care Reform." *Milbank
Quarterly* 79, no. 1 (March 2001): 35–54. https://doi.org/10.1111/1468-
0009.00195.

Melo, Marcus C. "State Retreat, Governance, and Metropolitan Restructuring
in Brazil." *International Journal of Urban and Regional Research* 19, no. 3
(1995): 342–57. https://doi.org/10.1111/j.1468-2427.1995.tb00512.x.

Melosi, Martin V. *The Sanitary City: Environmental Services in Urban America
from Colonial Times to the Present.* University of Pittsburgh Press, 2008.

Messina, Jane P., Oliver J. Brady, Nick Golding, Moritz U. G. Kraemer, G. R.
William Wint, Sarah E. Ray, David M. Pigott, et al. "The Current and Future
Global Distribution and Population at Risk of Dengue." *Nature Microbiology*
4, no. 9 (September 2019): 1508–15. https://doi.org/10.1038/s41564-019-
0476-8.

Metcalf, Alida C. *Go-Betweens and the Colonization of Brazil, 1500–1600.*
University of Texas Press, 2005.

Meyerson, Ann. "The Changing Structure of Housing Finance in the United
States." *International Journal of Urban and Regional Research* 10, no. 4
(December 1, 1986): 465–97. https://doi.org/10.1111/j.1468-2427.1986
.tb00025.x.

Ministério de Saúde. 2018. "Zika Vírus." Retrieved September 5, 2018. http://
portalms.saude.gov.br/saude-de-a-z/zika-virus.

Minnesota Population Center. "National Historical Geographic Information
System: Version 2.0." University of Minnesota, 2011.

Mittermeier, Russell A., Patricio R. Gil, Michael Hoffmann, John Pilgrim,
Thomas Brooks, Christina G. Mittermeier, John Lamoreux, and Gustavo
A. B. da Fonseca. *Hotspots Revisited: Earth's Biologically Richest and Most
Endangered Terrestrial Ecoregions.* Conservation International, 2005.

Montagu, Dominic, Andrew Anglemyer, Mudita Tiwari, Katie Drasser, George
W. Rutherford, Tara H. Horvath, Gail E. Kennedy, Lisa Bero, Nirali Shah,
and Heather Kinlaw. "Private Versus Public Strategies for Health Service
Provision for Improving Health Outcomes in Resource Limited Settings: A
Systematic Review." eScholarship: UCSF, 2011. https://escholarship.org/uc
/item/2dk6p1wz.

Moore, Andrew, Christina Nelson, Claudia Molins, Paul Mead, and Martin
Schriefer. "Current Guidelines, Common Clinical Pitfalls, and Future Directions for Laboratory Diagnosis of Lyme Disease, United States." *Emerging
Infectious Diseases* 22, no. 7 (July 2016): 1169–77. https://doi.org/10.3201
/eid2207.151694.

Mora, Camilo, Tristan McKenzie, Isabella M. Gaw, Jacqueline M. Dean, Hannah von Hammerstein, Tabatha A. Knudson, Renee O. Setter, et al. "Over Half of Known Human Pathogenic Diseases Can Be Aggravated by Climate Change." *Nature Climate Change* 12, no. 9 (September 2022): 869–75. https://doi.org/10.1038/s41558-022-01426-1.

Moraes, Fernando T. "Justiça autoriza Oxitec a comercializar mosquito transgênico." *Folha de S.Paulo,* March 26, 2018, sec. Ciência. https://www1.folha .uol.com.br/ciencia/2018/03/justica-autoriza-oxitec-a-comercializar -mosquito-transgenico.shtml.

Moraes Figueiredo, Luiz Tadeu. "Human Urban Arboviruses Can Infect Wild Animals and Jump to Sylvatic Maintenance Cycles in South America." *Frontiers in Cellular and Infection Microbiology* 9 (2019). https://doi .org/10.3389/fcimb.2019.00259.

Moran, Emilio F. "Deforestation and Land Use in the Brazilian Amazon." *Human Ecology* 21, no. 1 (March 1, 1993): 1–21. https://doi.org/10.1007 /BF00890069.

Mugagga, F., V. Kakembo, and M. Buyinza. "A Characterisation of the Physical Properties of Soil and the Implications for Landslide Occurrence on the Slopes of Mount Elgon, Eastern Uganda." *Natural Hazards* 60, no. 3 (February 1, 2012): 1113–31. https://doi.org/10.1007/s11069-011-9896-3.

Muhumuza, William. *Credit and the Reduction of Poverty in Uganda: Structural Adjustment Reforms in Context.* Fountain, 2007.

Mukunda, Gautam. "What Both Bernie Sanders and Donald Trump Get Wrong About Wall Street." *Washington Post,* April 21, 2016. https://www .washingtonpost.com/news/in-theory/wp/2016/04/21/what-both-bernie -sanders-and-donald-trump-get-wrong-about-finance/.

Murphy, Tim. "Is Romney Using Lyme Disease to Win Swing State Votes?" *Mother Jones,* October 22, 2012. https://www.motherjones.com/politics/2012 /10/mitt-romney-virginia-lyme-disease-michael-farris/.

Nasdaq. "Precigen, Inc. Common Stock (PGEN) Charts: Advanced Charting." Accessed May 25, 2022. https://www.nasdaq.com/market-activity/stocks /pgcn/advanced-charting.

National Institute of Allergy and Infectious Diseases. "Chronic Lyme Disease." 2018. https://www.niaid.nih.gov/diseases-conditions/chronic-lyme-disease.

Navarro, Vicente. "Neoliberalism and Its Consequences: The World Health Situation Since Alma Ata." *Global Social Policy* 8, no. 2 (2008): 152–55. https://doi.org/10.1177/14680181080080020203.

Neafsey, Daniel E., Robert M. Waterhouse, Mohammad R. Abai, Sergey S. Aganezov, Max A. Alekseyev, James E. Allen, James Amon, et al. "Highly Evolvable Malaria Vectors: The Genomes of 16 Anopheles Mosquitoes." *Science* 347, no. 6217 (January 2, 2015). https://doi.org/10.1126/science.1258522.

Neely, Abigail H. "Internal Ecologies and the Limits of Local Biologies: A Political Ecology of Tuberculosis in the Time of AIDS." *Annals of the Association of American Geographers* 105, no. 4 (2015): 791–805. https://doi.org/10.1080/00045608.2015.1015097.

Neely, Abigail H. *Reimagining Social Medicine from the South*. Duke University Press, 2021.

Nelsen, Lita. "The Role of University Technology Transfer Operations in Assuring Access to Medicines and Vaccines in Developing Countries." *Yale Journal of Health Policy, Law, and Ethics* 3 (Summer 2003): 293–300.

Netherton, Nan, Donald M. Sweig, Janice Artemal, Patricia Hickin, and Patrick Reed. *Fairfax County, Virginia: A History*. Fairfax County Board of Supervisors, 1978.

Nocera, Joe. *A Piece of the Action: How the Middle Class Joined the Money Class*. Simon and Schuster, 2013.

Norris, Douglas E. "Mosquito-Borne Diseases as a Consequence of Land Use Change." *EcoHealth* 1, no. 1 (March 1, 2004): 19–24. https://doi.org/10.1007/s10393-004-0008-7.

Nuijten, Monique, Martijn Koster, and Pieter de Vries. "Regimes of Spatial Ordering in Brazil: Neoliberalism, Leftist Populism, and Modernist Aesthetics in Slum Upgrading in Recife." *Singapore Journal of Tropical Geography* 33, no. 2 (2012): 157–70. https://doi.org/10.1111/j.1467-9493.2012.00456.x.

Nyakaana, Laban M. A. "Agricultural Development Planning and Policy in Uganda." PhD dissertation, University of Nairobi, 1970. Research Archive, University of Nairobi. http://erepository.uonbi.ac.ke/handle/11295/27033.

New York Stock Exchange. "UnitedHealth Group Inc. (UNH)." NYSE. Accessed February 17, 2023. https://www.nyse.com/quote/XNYS:UNH.

O'Hare, Greg, and Michael Barke. "The Favelas of Rio de Janeiro: A Temporal and Spatial Analysis." *Geojournal* 56 (January 3, 2002): 225–40. https://doi.org/10.1023/A:1025134625932.

O'Keefe, Eric. "Farmer Bill Gates." *The Land Report*, January 11, 2021. https://landreport.com/farmer-bill-gates/.

Okuonzi, Sam Agatre. "Dying for Economic Growth? Evidence of a Flawed Economic Policy in Uganda." *Lancet* 364, no. 9445 (2004): 1632–37. https://doi.org/10.1016/S0140-6736(04)17320-0.

Oliveira, Gustavo de L. T. "New Actors in China-Brazil Financial Geography: Commercial Banks and Currency Exchange Against the 'Resource-Seeking Consensus.'" *Professional Geographer* 75, no. 4 (October 24, 2022): 537–47. https://doi.org/10.1080/00330124.2022.2124179.

Oliveira, Rogério Ribeiro de. "When the Shifting Agriculture Is Gone: Functionality of Atlantic Coastal Forest in Abandoned Farming Sites." *Boletim do Museu Paraense Emílio Goeldi: Ciências Humanas* 3 (August 2008): 213–26. https://doi.org/10.1590/S1981-81222008000200006.

Oloya, J. J. "Marketing Boards and Post-War Economic Development Policy in Uganda, 1945–1962." *Indian Journal of Agricultural Economics* 23, no. 1 (March 1968): 50–58.

Olson, Kenneth E., Luke Alphey, Jonathan O. Carlson, and Anthony A. James. "Genetic Approaches in *Aedes aegypti* for Control of Dengue: An Overview." In *Bridging Laboratory and Field Research for Genetic Control of Disease Vectors*, edited by Bart G. J. Knols and Christos Louis, 77–87. Wageningen UR Frontis 11. Springer Netherlands, 2006. https://doi.org/10.1007/978-1-4020-3801-3_7.

Ostfeld, Richard S. *Lyme Disease: The Ecology of a Complex System*. Oxford University Press, 2011.

Ostfeld, Richard S., and Felicia Keesing. "Effects of Host Diversity on Infectious Disease." *Annual Review of Ecology, Evolution, and Systematics* 43, no. 1 (December 1, 2012): 157–82. https://doi.org/10.1146/annurev-ecolsys-102710-145022.

Ostfeld, Richard S., Taal Levi, Felicia Keesing, Kelly Oggenfuss, and Charles D. Canham. "Tick-Borne Disease Risk in a Forest Food Web." *Ecology* 99, no. 7 (2018): 1562–73. https://doi.org/10.1002/ecy.2386.

Overseas Development Institute. "Annual Report, 1964." London, 1964. https://cdn.odi.org/media/documents/8089.pdf.

Oxford University Innovation. "Oxford Spinout Oxitec Sold to Intrexon Corporation for $160 Million." Oxford University Innovation, August 10, 2015. https://innovation.ox.ac.uk/news/oxford-spinout-oxitec-sold-to-intrexon-corporation-for-160-million/.

Oxitec Ltd. "Oxitec Announces Friendly™ *Aedes* Project Is Supported by 92.8% of Piracicaba's Population." November 30, 2016. https://www.oxitec.com/en/news/oxitec-announces-friendly-aedes-project-is-supported-by-928-of-piracicabas-population.

Oxitec Ltd. "Oxitec Announces Ground-Breaking Commercial Launch of Its Friendly™ *Aedes aegypti* Solution in Brazil." November 3, 2021. https://www.oxitec.com/en/news/oxitec-announces-ground-breaking-commercial-launch-of-its-friendly-aedes-aegypti-solution-in-brazil.

Oxitec Ltd. "Oxitec Launches Field Trial in Brazil for Next Generation Addition to Friendly™ Mosquitoes Platform." May 24, 2018. https://www.oxitec.com/en/news/oxitec-launches-field-trial-in-brazil-for-next-generation-addition-to-friendly-mosquitoes-platform.

Oxitec Ltd. "Oxitec Receives US$6.8 Million from Wellcome Trust to Advance Scale-Up of Friendly™ *Aedes aegypti* Technology." April 21, 2021. https://www.oxitec.com/en/news/oxitec-receives-us68-million-from-wellcome-trust-to-advance-scale-up-of-friendly-aedes-aegypti-technology.

Oxitec Ltd. "Oxitec's Commercial Launch in Brazil!" YouTube, posted November 10, 2021. https://www.youtube.com/watch?v=i2KtpvZMZOU.

Oxitec Ltd. "Oxitec Successfully Completes First Field Deployment of 2nd Generation Friendly™ *Aedes aegypti* Technology." June 3, 2019. https://www.oxitec.com/en/news/oxitec-successfully-completes-first-field-deployment-of-2nd-generation-friendly-aedes-aegypti-technology.

Oxitec Ltd. "Oxitec Transitioning Friendly™ Self-Limiting Mosquitoes to 2nd Generation Technology Platform, Paving Way to New Scalability, Performance, and Cost Breakthroughs." November 28, 2018. https://www.oxitec.com/en/news/oxitec-transitioning-friendly-self-limiting-mosquitoes-to-2nd-generation-technology-platform-paving-way-to-new-scalability-performance-and-cost-breakthroughs.

Ozorio de Almeida, Anna Luiza. *The Colonization of the Amazon.* University of Texas Press, 1992.

Pan American Health Association (PAHO). "Timeline: Emergence of Zika the Virus in the Americas." PAHO and World Health Organization, Americas Region. Accessed November 6, 2024. https://www.paho.org/en/timeline-emergence-zika-virus-americas.

Parks, Brad, Ammar Malik, and Alex Wooley. "Is Beijing a Predatory Lender? New Evidence from a Previously Undisclosed Loan Contract for the Entebbe International Airport Upgrading and Expansion Project." AidData, February 27, 2022. https://docs.aiddata.org/reports/Uganda-Entebbe-Airport-China-Eximbank.html.

Patz, Jonathan A., Diarmid Campbell-Lendrum, Tracey Holloway, and Jonathan A. Foley. "Impact of Regional Climate Change on Human Health." *Nature* 438, no. 7066 (November 2005): 310–17. https://doi.org/10.1038/nature04188.

Pawar, Pratik. "It Took 35 Years to Get a Malaria Vaccine: Why?" *Undark Magazine*, May 25, 2022. https://undark.org/2022/05/25/it-took-35-years-to-get-a-malaria-vaccine-why/.

Perlman, Janice. *Favela: Four Decades of Living on the Edge in Rio de Janeiro.* Oxford University Press, 2010.

Peters, David H., Anu Garg, Gerry Bloom, Damian G. Walker, William R. Brieger, and M. Hafizur Rahman. "Poverty and Access to Health Care in Developing Countries." *Annals of the New York Academy of Sciences* 1136 (2008): 161–71. https://doi.org/10.1196/annals.1425.011.

Pinto, Vitor Laerte, Jr., José Cerbino Neto, and Gerson Oliveira Penna. "The Evolution of the Federal Funding Policies for the Public Health Surveillance Component of Brazil's Unified Health System (SUS)." *Ciência e Saúde Coletiva* 19 (December 2014): 4841–49. https://doi.org/10.1590/1413-812320141912.05962013.

Pires-Alves, Fernando Antônio, Carlos Henrique Assunção Paiva, and Nísia Trindade Lima. "Na Baixada Fluminense, à Sombra da 'Esfinge do Rio': Lutas Populares e Políticas de Saúde Na Alvorada do SUS." *Ciência e Saúde*

*Coletiva* 23, no. 6 (June 2018): 1849–58. https://doi.org/10.1590/1413
-81232018236.05272018.

Polanyi, Karl. *The Great Transformation: The Political and Economic Origins of Our Time.* Beacon Press, 2001.

Ponte, Stefano. "Coffee Markets in East Africa: Local Responses to Global Challenges or Global Responses to Local Challenges?" *CDR Working Paper No. 01.5.* 1 (2001).

Portes, Alejandro. "Housing Policy, Urban Poverty, and the State: The Favelas of Rio de Janeiro, 1972–1976." *Latin American Research Review* 14, no. 2 (1979): 3–24.

Pou, Sovitj, Rolf W. Winter, Aaron Nilsen, Jane Xu Kelly, Yuexin Li, J. Stone Doggett, Erin W. Riscoe, Keith W. Wegmann, David J. Hinrichs, and Michael K. Riscoe. "Sontochin as a Guide to the Development of Drugs Against Chloroquine-Resistant Malaria." *Antimicrobial Agents and Chemotherapy* 56, no. 7 (July 2012): 3475–80. https://doi.org/10.1128/AAC.00100-12.

Powell, Jeffrey R. "Mosquito-Borne Human Viral Diseases: Why *Aedes aegypti*?" *American Journal of Tropical Medicine and Hygiene* 98, no. 6 (June 2018): 1563–65. https://doi.org/10.4269/ajtmh.17-0866.

Powell, Jeffrey R., Andrea Gloria-Soria, and Panayiota Kotsakiozi. "Recent History of *Aedes aegypti*: Vector Genomics and Epidemiology Records." *Bioscience* 68, no. 11 (November 1, 2018): 854–60. https://doi.org/10.1093/biosci/biy119.

Prager, Alicia, and Flávia Milhorance. "Cerrado: Agribusiness May Be Killing Brazil's 'Birthplace of Waters.'" *Mongabay*, March 19, 2018. https://news.mongabay.com/2018/03/cerrado-agribusiness-may-be-killing-brazils-birthplace-of-waters.

Prasad, Eswar. *Gaining Currency: The Rise of the Renminbi.* Oxford University Press, 2017.

Presidência da República, Casa Civil, Subchefia para Assuntos Jurídicos. "Lei No. 4.380, de 21 de Agosto de 1964." 1964. http://www.planalto.gov.br/ccivil_03/leis/l4380.htm.

Proudhon, Pierre-Joseph. *The Solution of the Social Problem* (excerpts). 1927. The Anarchist Library, posted March 31, 2012. https://theanarchistlibrary.org/library/pierre-joseph-proudhon-the-solution-of-the-social-problem.

Quark, Amy A. *Global Rivalries: Standards Wars and the Transnational Cotton Trade.* University of Chicago Press, 2013. https://press.uchicago.edu/ucp/books/book/chicago/G/bo15997106.html.

Quinn, Sarah. *American Bonds: How Credit Markets Shaped a Nation.* Princeton University Press, 2019.

Radford, Gail. *Modern Housing for America: Policy Struggles in the New Deal Era.* University of Chicago Press, 2008.

Ramsdell, Lea. "National Housing Policy and the Favela in Brazil." In *The Political Economy of Brazil: Public Policies of an Era in Transition,* edited by Lawrence S. Graham and Robert Hines Wilson, 164–85. University of Texas Press, 1990.

Reis-Castro, Luisa, and Kim Hendrickx. "Winged Promises: Exploring the Discourse on Transgenic Mosquitoes in Brazil." In "Biotechnology, Controversy, and Policy: Challenges of the Bioeconomy in Latin America," edited by Pierre Delvenne and Kim Hendrickx. Special issue, *Technology in Society* 35, no. 2 (May 1, 2013): 118–28. https://doi.org/10.1016/j.techsoc.2013.01.006.

Reischauer, Robert. "Statement of Robert D. Reischauer, Deputy Director, Congressional Budget Office, Before the Subcommittee on Oversight, Committee on Ways and Means, U.S. House of Representatives." Congressional Budget Office. June 27, 1979. https://www.cbo.gov/sites/default/files/96th-congress-1979-1980/reports/doc233.pdf.

Requiao, R. G. "Brazilian Agricultural Export Promotion Experience to Advance Agricultural Trade: Legal, Regulatory, and Operational Frameworks and Impact Assessment." United Nations Food and Agriculture Organization. April 2016. https://openknowledge.fao.org/server/api/core/bitstreams/548b147d-aef2-4d3f-9ece-c3e102e7b415/content.

Richards, Mark David. "The Debates over the Retrocession of the District of Columbia, 1801–2004." *Washington History* 16, no. 1 (2004): 55–82.

Ridde, Valéry, Ludovic Queuille, and Marame Ndour. "Nine Misconceptions About Free Healthcare in Sub-Saharan Africa." *Development Studies Research* 1, no. 1 (January 1, 2014): 54–63. https://doi.org/10.1080/21665095.2014.925785.

Robbins, Paul. *Lawn People: How Grasses, Weeds, and Chemicals Make Us Who We Are.* Temple University Press, 2012.

Robbins, Paul. *Political Ecology: A Critical Introduction.* John Wiley and Sons, 2011.

Robinson, Kim Stanley. *The Ministry for the Future.* Orbit, 2020.

Robinson, Mark Dennis. *The Market in Mind: How Financialization Is Shaping Neuroscience, Translational Medicine, and Innovation in Biotechnology.* MIT Press, 2019.

Rodgers, Sarah E., Christine P. Zolnik, and Thomas N. Mather. "Duration of Exposure to Suboptimal Atmospheric Moisture Affects Nymphal Black-legged Tick Survival." *Journal of Medical Entomology* 44, no. 2 (March 2007): 372–75. https://doi.org/10.1603/0022-2585(2007)44[372:doetsa]2.0.co;2.

Rollend, Lindsay, Durland Fish, and James E. Childs. "Transovarial Transmission of Borrelia Spirochetes by *Ixodes scapularis*: A Summary of the Literature and Recent Observations." *Ticks and Tick-Borne Diseases* 4, no. 1 (February 1, 2013): 46–51. https://doi.org/10.1016/j.ttbdis.2012.06.008.

Rosário Costa, Nilson do. "Municipalização do controle da doença foi equívoco que custou caro ao país." Escola Nacional de Saude Pública Sergio Arouca, 2011. http://www6.ensp.fiocruz.br/visa/?q=node/5521.

Rose, Noah H., Massamba Sylla, Athanase Badolo, Joel Lutomiah, Diego Ayala, Ogechukwu B. Aribodor, Nnenna Ibe, et al. "Climate and Urbanization Drive Mosquito Preference for Humans." *Current Biology* 30, no. 18 (September 21, 2020): 3570–79. https://doi.org/10.1016/j.cub.2020.06.092.

Roy, Victor. *Capitalizing a Cure: How Finance Controls the Price and Value of Medicines.* University of California Press, 2023.

Rozell, Mark J., and Clyde Wilcox. *God at the Grass Roots, 1996: The Christian Right in the American Elections.* Rowman and Littlefield, 1997.

Sackett, David L., William M. C. Rosenberg, J. A. Muir Gray, R. Brian Haynes, and W. Scott Richardson. "Evidence Based Medicine: What It Is and What It Isn't." *BMJ* 312, no. 7023 (January 13, 1996): 71–72. https://doi.org/10.1136/bmj.312.7023.71.

Sampat, Bhaven N., and Frank R. Lichtenberg. "What Are the Respective Roles of the Public and Private Sectors in Pharmaceutical Innovation?" *Health Affairs* 30, no. 2 (February 2011): 332–39. https://doi.org/10.1377/hlthaff.2009.0917.

Santosa, Ailiana, Stig Wall, Edward Fottrell, Ulf Högberg, and Peter Byass. "The Development and Experience of Epidemiological Transition Theory over Four Decades: A Systematic Review." *Global Health Action* 7, no. 1 (December 1, 2014): 23574. https://doi.org/10.3402/gha.v7.23574.

Savage, Bruce. "Spin-Out Fever: Spinning Out a University of Oxford Company and Comments on the Process in Other Universities." *Journal of Commercial Biotechnology* 12, no. 3 (2006): 213–19.

Sawyer, Donald, Beto Mesquita, Bruno Coutinho, Fábio Vax de Almeida, Isabel Figueiredo, Ivana Lamas, Ludivine E. Pereira, Luiz P. Pinto, Mauro O. Pires, and Thaís Kasecker. "Ecosystem Profile: Cerrado Biodiversity Hotspot." Critical Ecosystem Partnership Fund, February 2017. https://www.cepf.net/sites/default/files/cerrado-ecosystem-profile-en-revised-2017.pdf.

Schmidt, Charles W. "Sprawl: The New Manifest Destiny?" *Environmental Health Perspectives* 112, no. 11 (August 2004): A620–A627.

Schmidt, Kenneth A., and Richard S. Ostfeld. "Biodiversity and the Dilution Effect in Disease Ecology." *Ecology* 82, no. 3 (2001): 609–19. https://doi.org/10.1890/0012-9658(2001)082[0609:BATDEI]2.0.CO;2.

Schmink, Marianne, and Charles H. Wood, eds. *Frontier Expansion in Amazonia.* University of Florida Press, 1985.

Schulthess, Duane, Harry P. Bowen, Robert Popovian, Daniel Gassull, Augustine Zhang, and Joe Hammang. "The Relative Contributions of NIH and Private Sector Funding to the Approval of New Biopharmaceuticals."

*Therapeutic Innovation and Regulatory Science* 57, no. 1 (2023): 160–69. https://doi.org/10.1007/s43441-022-00451-8.

Scott, W. Richard, Martin Ruef, Peter J. Mendel, and Carol A. Caronna. *Institutional Change and Healthcare Organizations: From Professional Dominance to Managed Care.* University of Chicago Press, 2000.

Sears, Cynthia L., Thomas M. File, Barbara D. Alexander, Daniel P. McQuillen, Ann T. MacIntyre, Upton D. Allen, Jonathan A. Colasanti, et al. "Charting the Path Forward: Development, Goals, and Initiatives of the 2019 Infectious Diseases Society of America Strategic Plan." *Clinical Infectious Diseases: An Official Publication of the Infectious Diseases Society of America* 69, no. 12 (November 27, 2019): e1–7. https://doi.org/10.1093/cid/ciz1040.

Segal, Lewis M., and Daniel G. Sullivan. "Trends in Homeownership: Race, Demographics, and Income." *Economic Perspectives* 22 (1998): 53–70.

Sejjaaka, Samuel. "A Political and Economic History of Uganda, 1962–2002." In *International Businesses and the Challenges of Poverty in the Developing World: Case Studies on Global Responsibilities and Practices*, edited by Frederick Bird and Stewart W. Herman, 98–110. Palgrave Macmillan, 2004. https://doi.org/10.1057/9780230522503_6.

Servick, Kelly. "Brazil Will Release Billions of Lab-Grown Mosquitoes to Combat Infectious Disease; Will It Work?" *Science,* October 16, 2016. https://www.science.org/content/article/brazil-will-release-billions-lab-grown-mosquitoes-combat-infectious-disease-will-it.

Seton, Ernest Thompson. *Lives of Game Animals.* Vol. 3. Doubleday, Page and Company, 1927.

Seukep, Sara E., Korine N. Kolivras, Yili Hong, Jie Li, Stephen P. Prisley, James B. Campbell, David N. Gaines, and Randel L. Dymond. "An Examination of the Demographic and Environmental Variables Correlated with Lyme Disease Emergence in Virginia." *EcoHealth* 12, no. 4 (December 1, 2015): 634–44. https://doi.org/10.1007/s10393-015-1034-3.

Shah, Sonia. *The Fever: How Malaria Has Ruled Humankind for 500,000 Years.* Farrar, Straus and Giroux, 2010.

Shearer, Freya M., Catherine L. Moyes, David M. Pigott, Oliver J. Brady, Fatima Marinho, Aniruddha Deshpande, Joshua Longbottom, et al. "Global Yellow Fever Vaccination Coverage from 1970 to 2016: An Adjusted Retrospective Analysis." *Lancet Infectious Diseases* 17, no. 11 (November 1, 2017): 1209–17. https://doi.org/10.1016/S1473-3099(17)30419-X.

Shiller, Robert. *Finance and the Good Society.* Princeton University Press, 2013.

Shinyekwa, Isaac, Julius Kiiza, Eria Hisali, and Marios Obwona. "The Evolution of Industry in Uganda." In *Manufacturing Transformation: Comparative Studies of Industrial Development in Africa and Emerging Asia*, edited by Carol Newman et al., 191–210. Oxford University Press, 2016.

Sica de Campos, André, Sarah Hartley, Christiaan de Koning, Javier Lezaun, and Lea Velho. "Responsible Innovation and Political Accountability: Genetically Modified Mosquitoes in Brazil." *Journal of Responsible Innovation* 4, no. 1 (January 2, 2017): 5–23. https://doi.org/10.1080/23299460.2017.1326257.

Sikkink, Kathryn A. *Ideas and Institutions: Developmentalism in Brazil and Argentina.* Cornell University Press, 1991.

Sistema Nacional de Informações sobre Saneamento (SNIS). "Série Histórica." Accessed November 5, 2024. http://app4.mdr.gov.br/serieHistorica/.

Siu, Albert L., Kirsten Bibbins-Domingo, and David Grossman. "Evidence-Based Clinical Prevention in the Era of the Patient Protection and Affordable Care Act: The Role of the US Preventive Services Task Force." *Journal of the American Medical Association* 314, no. 19 (November 17, 2015): 2021–22. https://doi.org/10.1001/jama.2015.13154.

Siviter, Harry, and Felicity Muth. "Do Novel Insecticides Pose a Threat to Beneficial Insects?" *Proceedings of the Royal Society B: Biological Sciences* 287, no. 1935 (September 30, 2020): 20201265. https://doi.org/10.1098/rspb.2020.1265.

Skidmore, Thomas E. *Brazil: Five Centuries of Change.* Oxford University Press, 1999.

Snyder, Robert E., Claire E. Boone, Claudete A. Araújo Cardoso, Fabio Aguiar-Alves, Felipe P. G. Neves, and Lee W. Riley. "Zika: A Scourge in Urban Slums." *PLOS Neglected Tropical Diseases* 11, no. 3 (March 23, 2017): e0005287. https://doi.org/10.1371/journal.pntd.0005287.

Sokol, Joshua. "The Worst Animal in the World." *The Atlantic,* August 20, 2020. https://www.theatlantic.com/health/archive/2020/08/how-aedes-aegypti-mosquito-took-over-world/615328/.

Solórzano, Alexandro, Lucas Santa Cruz de Assis Brasil, and Rogério Ribeiro de Oliveira. "The Atlantic Forest Ecological History: From Pre-colonial Times to the Anthropocene." In *The Atlantic Forest: History, Biodiversity, Threats, and Opportunities of the Mega-diverse Forest,* edited by Marcia C. M. Marques and Carlos F. V. Grelle, 25–44. Springer International, 2021. https://doi.org/10.1007/978-3-030-55322-7_2.

Soper, Fred L. "The Newer Epidemiology of Yellow Fever." *American Journal of Public Health and the Nations Health* 27, no. 1 (January 1937): 1–14. https://doi.org/10.2105/AJPH.27.1.1.

Souza, Flávio de, and Roger Zetter. "Urban Land Tenure in Brazil: From Centralized State to Market Processes of Housing Land Delivery." In *Market Economy and Urban Change,* edited by Roger Zetter and Mohammed Hamza, 163–84. Routledge, 2004.

Souza Salles, Tiago, Thayane da Encarnação Sá-Guimarães, Evelyn Seam Lima de Alvarenga, Victor Guimarães-Ribeiro, Marcelo Damião Ferreira de

Meneses, Patricia Faria de Castro-Salles, Carlucio Rocha dos Santos, et al. "History, Epidemiology, and Diagnostics of Dengue in the American and Brazilian Contexts: A Review." *Parasites and Vectors* 11, no. 1 (April 24, 2018): 264. https://doi.org/10.1186/s13071-018-2830-8.

Specter, Michael. "Mitt Romney versus Lyme Disease and Science." *New Yorker*, October 1, 2012. https://www.newyorker.com/news/news-desk/mitt-romney-versus-lyme-disease-and-science.

Spielman, A. "The Emergence of Lyme Disease and Human Babesiosis in a Changing Environment." *Annals of the New York Academy of Sciences* 740 (December 15, 1994): 146–56. https://doi.org/10.1111/j.1749-6632.1994 .tb19865.x.

Stapleton, Darwin H. "Lessons of History? Anti-malaria Strategies of the International Health Board and the Rockefeller Foundation from the 1920s to the Era of DDT." *Public Health Reports* 119, no. 2 (2004): 206–15.

Steere, A. C., S. E. Malawista, J. A. Hardin, S. Ruddy, W. Askenase, and W. A. Andiman. "Erythema Chronicum Migrans and Lyme Arthritis: The Enlarging Clinical Spectrum." *Annals of Internal Medicine* 86, no. 6 (June 1977): 685–98. https://doi.org/10.7326/0003-4819-86-6-685.

Stepan, Nancy. *Beginnings of Brazilian Science: Oswaldo Cruz, Medical Research, and Policy, 1890–1920.* Science History, 1976.

Strassburg, Bernardo B. N., Thomas Brooks, Rafael Feltran-Barbieri, Alvaro Iribarrem, Renato Crouzeilles, Rafael Loyola, Agnieszka E. Latawiec, et al. "Moment of Truth for the Cerrado Hotspot." *Nature Ecology and Evolution* 1, no. 4 (2017): 0099. https://doi.org/10.1038/s41559-017-0099.

Sweig, Donald M. "The Importation of African Slaves to the Potomac River, 1732–1772." *William and Mary Quarterly* 42, no. 4 (1985): 507–24. https://doi.org/10.2307/1919032.

Talbot, John M. *Grounds for Agreement: The Political Economy of the Coffee Commodity Chain.* Rowman and Littlefield, 2004.

Talbot, John M. "Information, Finance, and the New International Inequality: The Case of Coffee." *Journal of World-Systems Research*, August 26, 2002, 215–50. https://doi.org/10.5195/jwsr.2002.269.

Talisuna, Ambrose O., Abdisalan M. Noor, Albert P. Okui, and Robert W. Snow. "The Past, Present, and Future Use of Epidemiological Intelligence to Plan Malaria Vector Control and Parasite Prevention in Uganda." *Malaria Journal* 14, no. 1 (April 15, 2015): 158. https://doi.org/10.1186/s12936 -015-0677-4.

Teixeira, Maria Glória, Maria da Conceição N. Costa, Florisneide Barreto, and Maurício Lima Barreto. "Dengue: Twenty-Five Years Since Reemergence in Brazil." *Cadernos de Saúde Pública* 25 (2009): S7–18. https://doi.org/10.1590 /S0102-311X2009001300002.

Tett, Gillian. "We Should All Be Worried About the 'Financialisation' of Our World." *Financial Times,* May 24, 2023.

Thomas, D. D., C. A. Donnelly, R. J. Wood, and L. S. Alphey. "Insect Population Control Using a Dominant, Repressible, Lethal Genetic System." *Science* 287, no. 5462 (March 31, 2000): 2474–76. https://doi.org/10.1126 /science.287.5462.2474.

Thompson, Jordan L., and Karen E Francl-Powers. "A Brief History of Terrestrial Game Species Management in Virginia: 1900–Present," no. 41 (2013): 8.

Timmermans, Stefan, and Marc Berg. *The Gold Standard: The Challenge of Evidence-Based Medicine.* Temple University Press, 2010.

Troeger, Christopher. "Just How Do Deaths Due to COVID-19 Stack Up?" Think Global Health, Council on Foreign Relations, February 15, 2023. https://www.thinkglobalhealth.org/article/just-how-do-deaths-due-covid-19-stack.

Turshen, Meredeth. *Privatizing Health Services in Africa.* Rutgers University Press, 1999.

Turyahabwe, N., and A. Y. Banana. "An Overview of History and Development of Forest Policy and Legislation in Uganda." *International Forestry Review* 10, no. 4 (December 2008): 641–56. https://doi.org/10.1505 /ifor.10.4.641.

Uganda Coffee Development Authority. "Coffee Exports from 1964–2021." 2024. https://ugandacoffee.go.ug/resource-center/statistics.

Uganda Coffee Development Authority. "Uganda Coffee Development Authority Annual Report 2018–2019." 2019. https://ugandacoffee.go.ug/sites/default /files/2022-03/UCDA%20Annual%20Report_2018-2019.pdf.

Uganda Coffee Development Authority. "Uganda Coffee Development Authority Annual Report 2019–2020." 2020. https://ugandacoffee.go.ug/sites/default /files/2022-03/UCDA%20Annual%20Report_2019-2020_0.pdf.

Uganda Protectorate. *Annual Medical and Sanitary Report for the Year 1912.* Waterlow & Sons Limited, 1913. Wellcome Collection. https:// wellcomecollection.org/works/gvju7vkp/items.

Uganda Protectorate. *Annual Report of the Medical Department for the Year Ended 31st December, 1939.* Entebbe, Uganda: Government Printer, 1940. Wellcome Collection. https://wellcomecollection.org/works/eq7d6sg2 /items.

Uganda Protectorate. *Annual Report of the Ministry of Health for the Year 1959 and the Six Months from January to June, 1960.* Entebbe, Uganda: Government Printer, 1960. Wellcome Collection. https://wellcomecollection.org /works/v9vjhkry.

UNAIDS. "Global HIV & AIDS Statistics—Fact Sheet." Accessed September 21, 2023. https://www.unaids.org/en/resources/fact-sheet.

United States Census Bureau. "Housing Vacancies and Homeownership."
    Accessed November 5, 2024. https://www.census.gov/housing/hvs/data
    /histtabs.html.
United States Department of Agriculture Foreign Agricultural Service.
    "Commodity Intelligence Report." March 30, 2007. https://ipad.fas.usda
    .gov/highlights/2007/03/brazil_soybean_30mar2007/.
United States Department of Agriculture Foreign Agricultural Service. "World
    Agricultural Production." July 2018. https://downloads.usda.library.cornell
    .edu/usda-esmis/files/5q47rn72z/47429d90h/4q77fv87j/World_Ag_
    Production_July2018.pdf.
United States Department of Agriculture National Agricultural Statistics
    Service. "Land Values 2018 Summary 08/02/2018," 2018. https://www.nass
    .usda.gov/Publications/Todays_Reports/reports/land0818.pdf.
United States Department of Housing and Urban Development. "Survey of
    Mortgage Lending Activity—Mortgage Originations, 1–4 Family Units by
    Lender Type." HUD User, 1997. https://www.huduser.gov/periodicals
    /ushmc/winter97/histdat5.html.
United States Environmental Protection Agency. "Climate Change Indicators:
    Lyme Disease." EPA, July 1, 2016. https://www.epa.gov/climate-indicators
    /climate-change-indicators-lyme-disease.
United States Patent and Trademark Office. "Trademark Status and Document
    Retrieval (TSDR)." Accessed November 6, 2024. https://tsdr.uspto.gov/#case
    Number=87221245&caseSearchType=US_APPLICATION&caseType=DEF
    AULT&searchType=statusSearch.
United States Security and Exchange Commission. "Bill & Melinda Gates
    Foundation Form 13F." Accessed December 13, 2022. https://www.sec.gov
    /edgar/browse/?CIK=0001166559.
Vail, Stephen G., and Gary Smith. "Vertical Movement and Posture of Black-
    legged Tick (Acari: Ixodidae) Nymphs as a Function of Temperature and
    Relative Humidity in Laboratory Experiments." *Journal of Medical Ento-
    mology* 39, no. 6 (November 1, 2002): 842–46. https://doi.org/10.1603
    /0022-2585-39.6.842.
Valença, Márcio M. "The Inevitable Crisis of the Brazilian Housing Finance
    System." *Urban Studies* 29, no. 1 (February 1, 1992): 39–56. https://doi
    .org/10.1080/00420989220080041.
Valentine, Matthew John, Courtney Cuin Murdock, and Patrick John Kelly.
    "Sylvatic Cycles of Arboviruses in Non-Human Primates." *Parasites and
    Vectors* 12, no. 1 (October 2, 2019): 463. https://doi.org/10.1186/s13071-
    019-3732-0.
Valladares, Licia, and Ademir Figueiredo. "Housing in Brazil: An Introduction
    to Recent Literature." *Bulletin of Latin American Research* 2, no. 2 (1983):
    69–91. https://doi.org/10.2307/3338100.

Van Zee, Janice, Joseph F. Piesman, Andrias Hojgaard, and William Cormack Black IV. "Nuclear Markers Reveal Predominantly North to South Gene Flow in *Ixodes scapularis*, the Tick Vector of the Lyme Disease Spirochete." *PLoS ONE* 10, no. 11 (November 4, 2015). https://doi.org/10.1371/journal .pone.0139630.

Virginia Department of Game and Inland Fisheries. "Virginia Deer Management Plan 2015–2024." Virginia Department of Wildlife Resources, October 2015. https://dwr.virginia.gov/wildlife/deer/management-plan/.

Virginia Department of Health. "Tables of Selected Reportable Diseases in Virginia by Year of Report—Surveillance and Investigation." VDH, 2020. https://www.vdh.virginia.gov/surveillance-and-investigation/virginia-reportable-disease-surveillance-data/tables-of-selected-reportable-diseases-in-virginia-by-year-of-report/.

*Wall Street Journal.* "Metropolitan Life Sells Its Studebaker Convertible Preferred." September 28, 1959.

*Wall Street Journal.* "Studebaker Picks New Chief: Stepup in Mergers Likely." September 6, 1960.

Wallace, Robert G. *Dead Epidemiologists: On the Origins of COVID-19.* New York University Press, 2020.

Wallace, Robert G., Luke Bergmann, Richard Kock, Marius Gilbert, Lenny Hogerwerf, Rodrick Wallace, and Mollie Holmberg. "The Dawn of Structural One Health: A New Science Tracking Disease Emergence along Circuits of Capital." *Social Science and Medicine* 129 (March 1, 2015): 68–77. https:// doi.org/10.1016/j.socscimed.2014.09.047.

Wallace, Rodrick, Luis Fernando Chaves, Luke R. Bergmann, Constância Ayres, Lenny Hogerwerf, Richard Kock, and Robert G. Wallace. *Clear-Cutting Disease Control: Capital-Led Deforestation, Public Health Austerity, and Vector-Borne Infection.* Springer, 2018.

Walter, Katharine, Giovanna Carpi, Adalgisa Caccone, and Maria Diuk-Wasser. "Genomic Insights into the Ancient Spread of Lyme Disease across North America." *Nature Ecology and Evolution* 1 (August 28, 2017). https://doi .org/10.1038/s41559-017-0282-8.

Watira, W. "Bududa District Local Government Five-Year District Development Plan." Uganda National Planning Authority, 2011. http://www.npa.ug /wp-content/themes/npatheme/documents/East/Bududa%20DDP.pdf.

Wedig, Karin. *Cooperatives, the State, and Corporate Power in African Export Agriculture: The Case of Uganda's Coffee Sector.* New Political Economy 20. Routledge, 2019.

Weiss, Robin A., and Anthony J. McMichael. "Social and Environmental Risk Factors in the Emergence of Infectious Diseases." Supplement, *Nature Medicine* 10, no. S12 (December 2004): S70–76. https://doi.org/10.1038 /nm1150.

Wiegratz, Jörg. *Neoliberal Moral Economy: Capitalism, Socio-Cultural Change, and Fraud in Uganda.* Rowman & Littlefield, 2016.

Wilkinson, John, Bastiaan Reydon, and Alberto Di Sabbato. "Concentration and Foreign Ownership of Land in Brazil in the Context of Global Land Grabbing." *Canadian Journal of Development Studies / Revue canadienne d'études du développement* 33, no. 4 (2012): 417–38. https://doi.org.10.1080 .02255189.2012.746651.

Williams, Steven C. "Nationalism and Public Health: The Convergence of Rockefeller Foundation Technique and Brazilian Federal Authority During the Time of Yellow Fever, 1925–1930." In *Missionaries of Science: The Rockefeller Foundation and Latin America,* edited by Marcos Cueto, 23–51. Indiana University Press, 1994.

Willis, Evan, and Kevin Neil White. "Evidence Based Medicine, the Medical Profession, and Health Policy." In *Evidence-Based Health Policy: Problems and Possibilities,* edited by Vivian Lin and Brendan Gibson. Oxford University Press, 2003.

Windel, Aaron. *Cooperative Rule.* University of California Press, 2021.

Wolf, Martin. "Why Rigged Capitalism Is Damaging Liberal Democracy." *Financial Times,* September 18, 2019.

Wolf, Martin. "Wipe Out Rentiers with Cheap Money: Cautious Savers No Longer Serve a Useful Economic Purpose." *Financial Times,* May 6, 2014. https://www.ft.com/content/d442112e-d161-11e3-bdbb-00144feabdc0.

Wood, Chelsea L., and Kevin D. Lafferty. "Biodiversity and Disease: A Synthesis of Ecological Perspectives on Lyme Disease Transmission." *Trends in Ecology and Evolution* 28, no. 4 (April 1, 2013): 239–47. https://doi.org/10.1016 /j.tree.2012.10.011.

Woolhouse, Mark E. J., and Sonya Gowtage-Sequeria. "Host Range and Emerging and Reemerging Pathogens." *Emerging Infectious Diseases* 11, no. 12 (December 2005): 1842–47. https://doi.org/10.3201/eid1112.050997.

Woolhouse, Mark, Catriona Waugh, Meghan Rose Perry, and Harish Nair. "Global Disease Burden Due to Antibiotic Resistance: State of the Evidence." *Journal of Global Health* 6, no. 1 (June 5, 2016). https://doi.org/10.7189 /jogh.06.010306.

Wooten, James. "'The Most Glorious Story of Failure in the Business': The Studebaker-Packard Corporation and the Origins of ERISA." *Buffalo Law Review* 49, no. 2 (April 1, 2001): 683.

World Bank. "Agricultural Sector Adjustment." 1988. https://documents1. worldbank.org/curated/en/813611570647489652/pdf/Uganda-Agricu ltural-Sector-Adjustment.pdf.

World Bank. "Current Health Expenditure per Capita (Current US$)—Uganda: Data." 2022. https://data.worldbank.org/indicator/SH.XPD.CHEX.PC. CD?locations=UG.

World Bank. "External Debt Stocks, Total (DOD, Current US$)—Uganda."
2024. https://data.worldbank.org/indicator/DT.DOD.DECT.CD?end=2022&
locations=UG&start=1970.

World Bank. "Loan Agreement 2061 (Northwest Region Development Program—First Phase: Health Project) between Federative Republic of Brazil
and International Bank for Reconstruction and Development." December
15, 1981. https://documents1.worldbank.org/curated/
en/188111468230687148/pdf/Loan-2061-Brazil-Nw-Regional-Develop-
ment-Program-1st-Phase-Loan-Agreement.pdf.

World Bank. "Report No. 18142 Brazil: The Brazil Health System—Impact
Evaluation Report." June 30, 1998. https://documents1.worldbank.org
/curated/en/420601468744238297/pdf/Brazil-The-Brazil-health-system
.pdf.

World Bank. "WBG Finances—Country Details—Brazil." Finances One, World
Bank Group, 2023. https://financesapp.worldbank.org/countries/Brazil/.

World Bank. *World Development Report 1983*. Oxford University Press, 1983.
https://doi.org/10.1596/0-1952-0432-8.

World Health Organization. "Dengue and Severe Dengue." January 10, 2022.
https://www.who.int/news-room/fact-sheets/detail/
dengue-and-severe-dengue.

World Health Organization. *Good Procurement Practices for Artemisinin-Based
Antimalarial Medicines*. WHO Institutional Repository for Information
Sharing, 2010. https://apps.who.int/iris/handle/10665/44248.

World Health Organization. "One Year into the Zika Outbreak: How an
Obscure Disease Became a Global Health Emergency." WHO, 2016. http://
www.who.int/emergencies/zika-virus/articles/one-year-outbreak/en/.

World Health Organization. "Summary of Tuberculosis Data." Accessed
September 22, 2023. https://worldhealthorg.shinyapps.io/TBrief/?_inputs_
&sidebarCollapsed=true&sidebarItemExpanded=null.

World Health Organization. *World Malaria Report 2019*. 2019. https://apps
.who.int/iris/handle/10665/330011.

World Health Organization. *World Malaria Report 2021*. 2021. https://www
.who.int/teams/global-malaria-programme/reports/
world-malaria-report-2021.

World Health Organization. *World Malaria Report 2022*. 2022. https://www
.who.int/teams/global-malaria-programme/reports/
world-malaria-report-2022.

World Health Organization. "Yellow Fever." Accessed September 22, 2023.
https://www.who.int/news-room/fact-sheets/detail/yellow-fever.

Wormser, Gary P., Raymond J. Dattwyler, Eugene D. Shapiro, John J. Halperin,
Allen C. Steere, Mark S. Klempner, Peter J. Krause, et al. "The Clinical
Assessment, Treatment, and Prevention of Lyme Disease, Human

Granulocytic Anaplasmosis, and Babesiosis: Clinical Practice Guidelines by the Infectious Diseases Society of America." *Clinical Infectious Diseases* 43, no. 9 (November 1, 2006): 1089–1134. https://doi.org/10.1086/508667.

Wright, Eric R., and Brea L. Perry. "Medical Sociology and Health Services Research: Past Accomplishments and Future Policy Challenges." *Journal of Health and Social Behavior* 51, no. S1 (March 1, 2010): S107–S119. https://doi.org/10.1177/0022146510383504.

Worth, Katie, and Michelle Mizner. "Zika Uncontained." *Frontline*, August 10, 2016. https://www.pbs.org/wgbh/frontline/interactive/zika-water/.

Yamada, Tsuioshi. "The Cerrado of Brazil: A Success Story of Production on Acid Soils." *Soil Science and Plant Nutrition* 51, no. 5 (2005): 617–20. https://doi.org/10.1111/j.1747-0765.2005.tb00076.x.

Yang, H. M., M. L. G. Macoris, K. C. Galvani, M. T. M. Andrighetti, and D. M. V. Wanderley. "Assessing the Effects of Temperature on the Population of *Aedes aegypti*, the Vector of Dengue." *Epidemiology and Infection* 137, no. 8 (August 2009): 1188–1202. https://doi.org/10.1017/S0950268809002040.

Yearsley, Connor. "Artemisinin: A Nobel Prize–Winning Antimalarial from Traditional Chinese Medicine." *HerbalGram, The Journal of the American Botanical Council*, no. 110 (2016): 50–61.

Young, Crawford, Neal P. Sherman, and Tim H. Rose. *Cooperatives and Development: Agricultural Politics in Ghana and Uganda*. University of Wisconsin Press, 1981.

Zhang, Dayong, Min Hu, and Qiang Ji. "Financial Markets Under the Global Pandemic of COVID-19." *Finance Research Letters* 36 (October 2020): 101528. https://doi.org/10.1016/j.frl.2020.101528.

Zinszer, Kate, Kathryn Morrison, John S. Brownstein, Fatima Marinho, Alexandre F. Santos, and Elaine O. Nsoesie. "Reconstruction of Zika Virus Introduction in Brazil." *Emerging Infectious Diseases* 23, no. 1 (January 2017): 91–94. https://doi.org/10.3201/eid2301.161274.

# Index

Founded in 1893,
UNIVERSITY OF CALIFORNIA PRESS
publishes bold, progressive books and journals
on topics in the arts, humanities, social sciences,
and natural sciences—with a focus on social
justice issues—that inspire thought and action
among readers worldwide.

The UC PRESS FOUNDATION
raises funds to uphold the press's vital role
as an independent, nonprofit publisher, and
receives philanthropic support from a wide
range of individuals and institutions—and from
committed readers like you. To learn more, visit
ucpress.edu/supportus.

www.ingramcontent.com/pod-product-compliance
Lightning Source LLC
Chambersburg PA
CBHW020845270326
41928CB00006B/550